D0387404

REINVENTING GRAVITY

REINVENTING
GRAVITY

A PHYSICIST GOES BEYOND EINSTEIN

John W. Moffat

Smithsonian Books

COLLINS
An Imprint of HarperCollins Publishers

REINVENTING GRAVITY. Copyright © 2008 by John W. Moffat. All rights reserved. Printed in the United States of America. No part of this book may be used or reproduced in any manner whatsoever without written permission except in the case of brief quotations embodied in critical articles and reviews. For information, address HarperCollins Publishers, 10 East 53rd Street, New York, NY 10022.

All images appearing throughout the text are © Patricia Wynne, except Figures 6, 10, 12, 13, 14, 17, and 19.

HarperCollins books may be purchased for educational, business, or sales promotional use. For information, please write: Special Markets Department, HarperCollins Publishers, 10 East 53rd Street, New York, NY 10022.

FIRST EDITION

Designed by Daniel Lagin

Library of Congress Cataloging-in-Publication Data
Moffat, John W.
Reinventing gravity: a physicist goes beyond Einstein/John W. Moffat.
    p.   cm.
Includes bibliographical references.
ISBN 978-0-06-117088-1
1. Quantum gravity. 2. Astrophysics. 3. General relativity (Physics) I. Title.
QC178.M64 2008
530.14'3—dc22   2008013391

08 09 10 11 12   OV/RRD   10 9 8 7 6 5 4 3 2 1

QC
178
m64
2008

To Patricia

whose dedication and help
made this book possible

Contents

Introduction

A NEW GRAVITY THEORY

I n 1916, Einstein published his new theory of gravity called general relativity. In 1919, the theory was validated by the observation of the bending of light during a solar eclipse, as the sun's gravitational pull warped spacetime. Since then, there has been much speculation as to whether Einstein's theory of gravity is perfect and unchangeable, much like Michelangelo's David. Why would we want to modify Einstein's outstanding intellectual achievement?

Until recently, most physicists have considered Einstein's general relativity theory to be in perfect agreement with observational data. However, this is not necessarily true. Neither have the attempts succeeded to unify Einstein's gravitational theory with quantum mechanics, despite much effort: Many physicists consider the search for a successful quantum gravity theory the holy grail of modern physics. Moreover, there are some fundamentally unsatisfactory features of Einstein's theory, such as those related to the beginning of the universe and the collapse of stars under their own gravitational forces.

Finally, since the early 1980s, a growing amount of observational data has been accumulating that shows that Newtonian and Einstein gravity cannot describe the motion of the outermost stars and gas in galaxies correctly if only their visible mass is accounted for in the gravitational field equations.* There are much stronger gravitational forces being observed—causing the peripheral orbiting stars and gas to move faster—than are predicted by Newton's and Einstein's theories. There is now overwhelming evidence for stronger

---

* Einstein's gravity theory reduces to Newton's theory for weak gravitational fields and slowly moving bodies.

gravity in galaxies. To put this in perspective, consider that Einstein's correction to Mercury's orbit, which constituted a major test for general relativity theory in 1915, was tiny, representing only 43 arc seconds *per century*.* In contrast, the discrepancy between the rotational speeds of stars in the outermost parts of giant spiral galaxies and the predictions of the prevailing theories of gravity is enormous: The stars are moving at about *twice* the speed that they should be, according to Newtonian and Einstein gravity.

To save Einstein's and Newton's theories, many physicists and astronomers have postulated that there must exist a large amount of "dark matter" in galaxies and also clusters of galaxies that could strengthen the pull of gravity and lead to an agreement of the theories with the data. The clusters and superclusters of galaxies are the largest observed objects in the universe. Most physicists believe that without dark matter, the clusters of galaxies could not be stable objects, for the gravitational force determined by Einstein's and Newton's theories is not strong enough to hold the galaxies together in the cluster. This invisible and undetected dark matter neatly removes any need to modify Newton's and Einstein's gravitational theories. Invoking dark matter is a less radical, less scary alternative for most physicists than modifying the theories; indeed, for almost a century, Einstein's gravity theory has constituted the "standard model of gravity." Thus a consensus has formed among astronomers and physicists that dark matter really exists, even though all attempts to detect the particles that supposedly constitute the dark matter have so far failed.

Since 1998, this disturbing discrepancy between theory and observation in the galaxies and clusters of galaxies has been compounded by a discovery that has catapulted cosmology into a state of chaos. Two independent groups of astronomers in California and Australia have found that the expansion of the universe is actually accelerating, rather than slowing down, which is what one would expect. In order to explain this remarkable discovery, physicists have postulated that there must exist a mysterious form of "dark energy" that has negative pressure and therefore can act like antigravity. In contrast to the putative dark matter, it must be uniformly distributed throughout the universe. Like dark matter, though, it can only be detected through the action of gravity. Together with the dark matter, this means that about 96 percent of all the matter and energy comprising the universe is invisible! This remarkable conclusion is held to be true by the majority of physicists and astronomers today, and is an accepted part of the standard model of cosmology.

---

* In astronomy, an arc minute is an angle in the sky that is one-sixtieth of a degree, and an arc second is one-sixtieth of an arc minute.

Could it be, nevertheless, that Einstein's theory is wrong? Might it be necessary to modify it—to find a new theory of gravity that can explain both the stronger gravity and the apparent antigravity being observed today—rather than simply throwing in invisible things to make the standard model work?

Our view of the universe changed dramatically from the mechanistic universe of Newton when Einstein published his new gravity theory. Through Einstein, we understood that space and time were not absolute, that gravity was not a "force," but a characteristic of the geometry of "spacetime," and that matter and energy warped the geometry of spacetime. Yet Einstein himself was never satisfied with his revolutionary theory. He was always seeking a more complete theory.

After publishing the final form of general relativity in March 1916, Einstein attempted to construct a unified theory of gravitation and electromagnetism. At that time, these were the only known forces in nature. This led to a discouraging series of fruitless attempts over many years, as Einstein tried to repeat the outstanding success of his discovery of the law of gravitation. Einstein never did succeed in unifying the forces of nature, but a primary motivation for attempting this intellectual feat was his dissatisfaction with one feature of general relativity: At certain points in spacetime his equations developed singularities; that is, the solutions of the equations became infinitely large.

This disturbing feature has important consequences for the view of the universe depicted by Einstein's general relativity. At the birth of the universe, according to the generally accepted standard model of cosmology, the big bang starts in an infinitely small volume of space with matter of infinite density. Similarly, when a star collapses under its own gravitational forces, the density of the matter in the star becomes infinitely large as it collapses into an infinitely small volume. One of the predictions of general relativity is that when the mass of a star is greater than a certain critical value, and if it collapses under its own gravitational force, then it forms a black hole. No light can escape the black hole, and at its center lurks an infinitely dense singularity. Einstein was not in favor of the black hole solutions of his theory, because he considered them unphysical. The astronomer Sir Arthur Eddington, who was a champion of Einstein and led one of the first expeditions to observe the bending of light during a solar eclipse, was also adamantly opposed to the idea of stars collapsing into black holes. "I think there should be a law of Nature to prevent a star from behaving in this absurd way!" he declared in a speech to the Royal Astronomical Society.[1] Thus, finding a more general framework that would describe the structure of spacetime so that it did not contain unphysical singularities was one of the primary motivations for Einstein's attempts to create a more general unified theory.

When I began studying physics as a young student in Copenhagen in the early 1950s, outside the academic environment, my first objective was to understand Einstein's latest work on his unified theory. This was the first step on my lifelong path to develop a generalization of Einstein's gravitational theory. In his later years, Einstein worked on what he called the "nonsymmetric field theory." He claimed that this theory contained James Clerk Maxwell's equations for electromagnetic fields. This turned out to be false, and Einstein died in 1955 before he was able to complete this work.

In a series of letters between Einstein and myself when I was twenty years old and he was nearing the end of his life, we discussed his current work on unified field theory as well as how he viewed the discipline of physics at that time. In retrospect, it is clear to me, looking back over my own research during the past fifty years, that I have been following in Einstein's footsteps, tracking the course he would have taken if he were alive today.

The story I tell in this book is my quest for a new gravity theory. Like Einstein, I began by constructing a Nonsymmetric Gravitation Theory (NGT). In contrast to Einstein, however, I did not consider the nonsymmetric theory a unified theory of gravitation and electromagnetism, but a generalized theory of pure gravity. Over a period of thirty years, with successes and setbacks along the way, I developed NGT into ever simpler versions that eventually became my Modified Gravity Theory (MOG). In Parts I, II, and III of this book, I set the new ideas of MOG against the historical background of the discovery of gravity, Einstein's monumental contributions, and the development of modern cosmology. In Parts IV and V, I describe how I developed MOG and how it departs from Einstein's and Newton's theories, as well as from the consensus view of mainstream cosmologists, physicists, and astronomers. I also set MOG in the context of other alternative gravity theories, including strings and quantum gravity.

I have had two types of readers in mind during the writing of this book: the curious non-physicist who loves science, and the reader with a more technical background in physics. If you are the first type of reader, I encourage you to forge ahead through any sections that might seem difficult. There will be generalizations and comparisons coming that will enable you to grasp the significance of challenging passages and to see the emerging big picture. If you are the second type of reader, then the notes at the end of the book are designed for you, if you would like to know more about the mathematics and other technical details behind the topics.

MOG solves three of the most pressing problems in modern physics and cosmology. Because MOG has stronger gravity than the standard model, it does away entirely with the need for exotic dark matter. It also explains the origin of the dark energy. Moreover, MOG has no vexing singularities. Through the mathematics of the theory, MOG reveals the universe to be a different kind of place—perhaps even a more straightforward and sensible place—than we have been led to believe.

MOG contains a new force in nature, a fifth force similar in strength to gravity. It makes its presence known beyond the solar system, in the motions of stars in galaxies, clusters of galaxies, and in the large-scale structure of the universe—which is to say, in the entire universe. This fifth force appears as a new degree of freedom in the equations of MOG, and is one of the primary ingredients when we generalize Einstein's theory. Now we can say that there are five basic forces of Nature: gravitation, electromagnetism, the weak force that governs radioactivity, the strong force that binds the nucleus in an atom, and the new force. This new force changes the nature of the warping of the spacetime geometry in the presence of matter. It has a "charge" that is associated with matter, much as electric charge is responsible for the existence of electric and magnetic fields. However, in one representation of the theory, the fifth force can be thought of as a gravitational degree of freedom—part of the overall geometry or warping of spacetime.

Another important feature of the new theory is that Newton's gravitational constant is no longer a constant, but varies in space and time. Together with the fifth force, this varying element strengthens the pull of gravity in faraway galaxies and in clusters of galaxies. It also alters the geometry of spacetime in the expanding universe, and allows for an agreement with the satellite observations of the cosmic microwave background. MOG does away with the need for dark matter in cosmology, as well as in the galaxies and clusters of galaxies. It also alters Newton's inverse square law of gravity for weak gravitational fields.

MOG changes our view of what happens when a star collapses under its own gravitational forces. Because general relativity's prediction of a black hole can be changed in MOG, we can only say that the star collapses to a very dense object, which is not exactly black, but possibly "grey." Therefore, information such as some light can escape the "grey star." The dark energy, or vacuum energy, that is responsible for the acceleration of the universe can also play an important role in stabilizing astrophysical bodies against gravitational collapse. Thus MOG may dramatically change our view of black holes, one of the most exotic predictions of Einstein's gravitational theory.

Finally, if MOG turns out to be true, one of the most popular hypotheses in science may fall: The big bang theory may be incorrect as a description of the very early universe. Because of the smoothness of spacetime in MOG, there is no actual singular beginning to the universe, although there is a special time equal to zero (t = 0), as there is in the big bang theory. But in MOG, t = 0 is free of singularities. The universe at t = 0 is empty of matter, spacetime is flat, and the universe stands still. Because this state is unstable, eventually matter is created, gravity asserts itself, spacetime becomes curved and the universe expands. In contrast to the big bang scenario, the MOG universe is an eternal, dynamically evolving universe—which may have implications for philosophy and religion as well as astrophysics and cosmology.

Adopting a new gravity theory means changing a major scientific paradigm that has endured and served us well for many years. Newton's gravity theory, still correct within the solar system, including on the Earth, has been valid for more than three centuries. Einstein's general relativity has stood the test of time for almost a century. Physicists tend to be conservative when faced with the possibility of overthrowing a paradigm, and that is how it should be. Only when a mass of accumulating new data provides overwhelming observational evidence that a revolution is necessary are most scientists willing to support a new theory.

It has become fashionable in modern physics to construct mathematical castles in the air, with little or no relation to reality, little or no hope of ever finding data to verify those theories. Yet fitting the observational data is a driving force in my work on MOG. Throughout this book, I stress the importance of finding data that can verify or falsify the theory, for I believe strongly that a successful theory of nature must be grounded in observation.

Let me state the current situation very clearly. There are only two ways of explaining the wealth of observational data showing the surprisingly fast rotational speeds of stars in galaxies and the stability of clusters: Either dark matter exists and presumably will be found, and Newton's and Einstein's gravity theories will remain intact; or dark matter does not exist and we must find a new gravity theory.

Today it is possible that we are standing on the brink of a paradigm shift. It is likely that the consensus that dark matter exists will serve as the tipping point, and that someday the dark matter hypothesis will seem as embarrassing as the emperor's new clothes.

REINVENTING GRAVITY

Prologue

THE ELUSIVE PLANET VULCAN, A PARABLE

For a time in the late nineteenth century, the planet Vulcan was considered a reality—an observed planet. The great French mathematical astronomer Urbain Jean Joseph Le Verrier predicted the existence of the new planet in 1859 based on astronomical observations and mathematical calculations using Newton's gravitational equations. Since the new planet would now be the closest planet to the sun, closer than its hot neighbor Mercury, Le Verrier christened it "Vulcan" after the Roman god of fire and iron. (The Greeks had named the planets after the gods of Olympus, and later the Romans translated them into the names by which the inner planets are known today: Mercury, Venus, Mars, Jupiter, and Saturn.)

Le Verrier had discovered an anomaly in the orbit of Mercury. Like all planets, Mercury traced an ellipse in its orbit around the sun, and the position of its closest approach to the sun, called the "perihelion," advanced with successive revolutions. The pattern of the planet's orbit thus looked like a complex rosette shape over time. Le Verrier had already taken into account the effects of the other planets on Mercury's orbit, calculating the precession accurately according to Newton's celestial mechanics. However, there was a small discrepancy between the Newtonian prediction for the perihelion precession and contemporary astronomical observations. According to Le Verrier, this could only be explained by the gravitational pull of an as-yet-unseen planet or perhaps a ring of dark, unseen asteroids around the sun. A hunt soon ensued, as astronomers around the world vied to become the first to observe Vulcan.

Le Verrier had already had resounding success in predicting the existence of an unknown planet. In 1846, he had published his calculations of the wayward movements in the orbital motion of Uranus, the planet discovered by William Herschel in 1781. Uranus's orbit had long been a problem for astronomers, and several of them claimed that there must be an error in Newton's universal law of attraction. From his calculations using Newton's gravity theory, Le Verrier concluded that there must be another planet orbiting the sun on the far side of Uranus. He christened it "Neptune" after the god of the sea. He even predicted the approximate coordinates of the planet's position in the solar system. In that same year, on the instructions of Le Verrier, the German astronomer Johann Gottfried Galle discovered the planet Neptune at right ascension 21 hours, 53 minutes and 25.84 seconds, very close to where Le Verrier had predicted it would be.

This was a great triumph for Le Verrier, and his fame spread. He became one of the most influential astronomers in the world, and in 1854 succeeded the renowned doyen of French astronomers, François Jean Dominique Arago, as director of the Paris Observatory. In 1859, Le Verrier fully expected Vulcan to eventually be identified by astronomers' telescopes, just as Neptune had been thirteen years earlier.

From the vantage point of early-twenty-first-century physics and cosmology, we can use the modern term "dark matter" to characterize the predictions of Neptune and Vulcan. That is, gravitational anomalies in the orbits of two known planets implied the existence of unknown, unseen, or "dark" objects nearby. Those unseen bodies would be responsible for the extra gravitational force exerted on the known planets' orbits. Their discovery would explain the anomalies in the observational data and would yet again demonstrate the correctness of Sir Isaac Newton's gravitational theory.

In March 1859, in Orgères-en-Beauce, France, a small community about twenty miles north of Orléans in the Loire district, an amateur astronomer and physician, Edmond Modeste Lescarbault, observed a small black dot crossing the face of the sun from the small observatory adjacent to his clinic. Having heard of Le Verrier's prediction of the planet Vulcan, he excitedly identified the black dot as the missing planet. In December 1859, Lescarbault wrote to Le Verrier claiming that he had verified the existence of the dark planet Vulcan. Together with an assistant, the busy imperial astronomer paid a visit to Lescarbault in Orgères, completing the final part of the journey on foot through the countryside from the railway station.

The imperious Le Verrier was at first dismissive of Lescarbault's claim, and cross-examined him for an hour. Nevertheless, he was strongly motivated to accept the rural physician's discovery, and left Orgères convinced that the dark planet had been found. The news of the discovery took Paris by storm, and Lescarbault became famous overnight. In 1860 the emperor Napoléon III conferred the Légion d'Honneur on him, and the Royal Astronomical Society in England lavished praise on him. Thus the village physician and amateur astronomer had solved one of the greatest mysteries of nineteenth-century astronomy. He had found the "dark matter" predicted by Newton's gravitational equations.

Unfortunately for Lescarbault and Le Verrier, the story does not end here. The accolades eventually turned to accusations. Subsequent investigators never found Vulcan, even though several solar eclipse expeditions were mounted in various parts of the world to find the best vantage point from which to observe Vulcan traversing the face of the sun. At the July 29, 1878, eclipse of the sun, the last total solar eclipse observable from the United States in the nineteenth century, the astronomer Craig Watson claimed to have observed Vulcan from Rawlins, Wyoming, in Indian country. Watson devoted the rest of his life to defending this claim, and near the end of his life he constructed an underground observatory that he hoped would vindicate him.

As it turned out, the dark planet supposedly seen by Watson was not big enough to explain the anomalous precession of Mercury, so various astronomers proposed that other dark matter planets must be waiting to be discovered too. Lewis Swift, an amateur astronomer from Rochester, New York, also claimed to have observed the dark planet by independent observations at the time of the solar eclipse, thereby confirming Watson's discovery. However, Christian Peters of Germany was highly skeptical of the whole Vulcan enterprise, and publicly disputed the claims of both Watson and Swift, becoming an archenemy of Watson in the process.

With no widely accepted verification of Lescarbault's original discovery, it wasn't long before scientific papers began appearing in astronomical journals disputing the Vulcan discovery. Articles even appeared in the English and French newspapers discussing these disclaimers by important astronomers. Meetings were held at the astronomical societies in London and Paris, with papers presented for and against the existence and discovery of Vulcan. Although Le Verrier continued to maintain that his prediction was correct and Vulcan would eventually be found, this nineteenth-century "dark matter" problem subsided into an unresolved, sleeper issue for sixty-five years. Le Verrier and most of the astronomers who had participated in the hunt died without knowing the ending to the Vulcan story.

The problem of the anomalous perihelion advance of Mercury remained. The German astronomer Hugo von Seeliger proposed his zodiacal light theory in 1906. He suggested that small, ellipsoidal concentrations of dark matter particles existed near the sun, and were responsible for the anomalous perturbation of Mercury's orbit. Zodiacal light is light reflected from dust particles left by comets and asteroids in the solar system. Even the famous French mathematician Henri Poincaré postulated rings of dark matter particles around the sun, and the celebrated American astronomer Simon Newcomb supported these suggestions.

The decades-long battle over the elusive Vulcan only ended in 1916 when Einstein published his general theory of relativity. In 1915, when he was still developing his new gravity theory, Einstein performed a calculation investigating the anomalous advance of the perihelion of Mercury, which a half-century earlier had created such excitement and the prediction of a new planet. He discovered that when he inserted the mass of the sun and Newton's gravitational constant into his equations, his emerging theory of general relativity correctly predicted the strange precession of Mercury's orbit. Thus, Le Verrier's Vulcan was discarded and the anomalous orbit of Mercury turned out to be caused by a predictable warping of the geometry of spacetime near the sun. Einstein's new theory of gravity was something that the nineteenth-century astronomers could never have imagined. A new planet or dark matter particles ringing the sun were no longer needed to explain the Mercury anomaly. Thus the dark and nonexistent planet Vulcan served as a watershed in the history of gravitation theory and celestial mechanics. Einstein's gravity theory precipitated a revolutionary change in our understanding of space and time.

Einstein's calculations were in startling agreement with more than a century of observations of Mercury's orbit. Einstein discovered that Mercury should precess faster than the predicted Newtonian speed by the tiny amount of 0.1 arc seconds for each orbital revolution, amounting to 43 arc seconds per century, very close to the observed value. With this first success, Einstein realized that he was definitely on the right track: He had a robust new gravity theory that would eventually overthrow Newton's.

Thus the Vulcan "dark matter" problem was resolved not by the detection of a dark planet or dark matter particles but by modifying the laws of gravitation. Le Verrier turned out to be right with his prediction of the discovery of Neptune, and wrong with the prediction of Vulcan.

We are confronted with a surprisingly similar situation today in physics and cosmology. Astronomical observations of the motion of stars in galaxies, and the motion of galaxies within clusters of galaxies, do not agree with either Newton's or Einstein's gravitation equations when only the visible matter making up the stars, galaxies, and gas is taken into account. Much stronger gravity and acceleration are actually being observed in faraway stars and galaxies than one would expect from Newton's and Einstein's gravitational theories. In order to fit the theories to the observational data for galaxies and clusters of galaxies, many scientists have proposed that a form of unseen "dark matter" exists, which would account for the strong effects of gravity that are being observed. That is, there seems to be much more matter "out there" than we have so far been able to see. The hunt is now on to observe, and to actually find, this missing dark matter. It is claimed that 96 percent of the matter and energy in the universe is invisible, or dark. Almost 30 percent of the total matter-energy budget is said to be composed of so-called cold dark matter and almost 70 percent of "dark energy," leaving only about 4 percent as visible matter in the form of the atoms that make up the stars, planets, interstellar dust, and ourselves. Such is the degree of the discrepancy between theory and observations today.

How will the modern dark matter story end? Like Neptune or like Vulcan? Will dark matter be found or not? Are the observed motions of stars in galaxies caused by dark matter that, when added to Newton's and Einstein's laws of gravity, speeds up the motion of stars? Or does dark matter not exist, and once again in the history of science we must face modifying the laws of gravity? Will dark matter tip us into another revolution in our understanding of gravity?

PART 1

DISCOVERING AND REINVENTING GRAVITY

Chapter 1

THE GREEKS TO NEWTON

I s there any phenomenon in physics as obvious as the force of gravity? Gravity keeps the planets in their orbits around the sun and holds stars together in galaxies. It prevents us from floating off the Earth, makes acorns and apples fall down from trees, and brings arrows, balls, and bullets to the ground in a curved path.

Yet gravity is so embedded in our environment that many thousands of years passed before humans even perceived gravity and gave it a name. In fact, the everyday evidence of gravity is extremely difficult to "see" when one lives on one planet, without traveling to another for comparison. The little prince in Saint-Exupéry's tale would have formed a vastly different idea of gravity from living only on his small asteroid. To early human beings, just as to most of us today, the behavior of falling objects is a practical experience taken for granted rather than an example of a universal force. Because the Earth is large and we only experience gravity from the effects of Earth, and not some other object, we tend to think of gravity as "down," without realizing that gravity is a property of bodies in general. In contrast, electromagnetism is a much more obvious force. We see it in lightning and magnets and feel it in static electricity. But it took many centuries to discover gravity, and some promising ideas along the way turned out to be completely wrong. It wasn't until the late seventeenth century that Isaac Newton recognized that the same force of attraction or "togetherness" that ruled on the Earth also bound objects in the heavens. The paradigm shift that Newton wrought was in understanding gravity as a universal force.

The story of the discovery of gravity is also the story of astronomy, especially the evolving ideas about the solar system. Western science originated with the Greeks, whose model of an Earth-centered universe dominated scientific thought for almost 2,000 years. The Greek mind was abstract, fond of ideals and patterns, and slipped easily into Christianity's Earth- and human-centered theology. It took many centuries for thinkers such as Copernicus, Kepler, Galileo, and Newton to break with the enmeshed Platonic and Christian views of the universe, to turn astronomy and physics into sciences, and to develop the idea of gravity.

GREEK ASTRONOMY AND GRAVITY

Plato's most famous student was Aristotle (384–322 BC), whose system of thought formed the basis of Western science and medicine until the Renaissance. For Aristotle, four elements composed matter: Earth, Water, Air, and Fire. Earth, as the basest and heaviest of the elements, was at the center of the universe. Although Aristotle did not use the Greek equivalent of the word "gravity," he believed that people and objects did not fall off the Earth because they were held by the "heaviness" of Earth.

Plato had taught that nature's most perfect shapes were the circle, in two dimensions, and the sphere in three. Aristotle's cosmology in turn relied heavily on circles and spheres. Around Aristotle's Earth, several "crystalline spheres" revolved. First were the Earth-related spheres of Water, Air, and Fire. Spheres farther out contained the heavenly bodies that appeared to move around the Earth: the moon, sun, and the five planets known to the Greeks (Mercury, Venus, Mars, Jupiter, and Saturn). Beyond these was the sphere of the "fixed stars," while the final sphere was the dwelling place of God, the Prime Mover of all the spheres. This cosmology needed spheres for the heavenly bodies to move on because Aristotle believed that objects could move only when in contact with another moving object. The spheres were "crystalline" rather than translucent or opaque because the fixed stars had to be visible to observers on Earth through the other rotating spheres.

The ancient Greeks knew that the Earth was a sphere, but most astronomers pictured it as a static object, immovable at the center of the universe. This view stemmed from "common sense": In our everyday experience, barring earthquakes, we do not sense any motion of the Earth. Also, if the Earth moved through the heavens, the Greeks argued, we would observe stellar parallax. Parallax can easily be demonstrated by holding a finger up in front of one's face and closing first one eye and then the other; the finger appears to be moving from side to side relative to objects in the background. Similarly, if the Earth moved through the heavens, then the stars nearest to Earth in the stellar sphere would move relative to those more distant. Since this did not happen, Aristotle concluded that the Earth stayed still at the center of the universe.

Aristotle and his contemporaries did not conceive of the vast distances that actually separate objects in the universe, and believed that the "fixed stars" were thousands of times closer to Earth than they are. In fact, we *can* detect parallax in the nearer stars today, as the Earth moves around the sun, and more dramatically as the solar system moves around the Milky Way galaxy. Powerful telescopes take photographs of the same stars at different points in the Earth's or solar system's orbit, and differences in the stars' positions relative to background stars can be seen when comparing those photographs.

One prominent Greek astronomer, Aristarchus of Samos (310–230 BC), did propose a heliocentric universe. He had figured out that the sun was a great deal larger than the Earth, and it made more sense to him that a smaller object would orbit a larger one. Aristarchus concluded that the universe was much larger than most people believed, that the sun was at its center, and that neither the sun nor the sphere of fixed stars moved. Aristarchus correctly placed the moon in orbit around the Earth, and all the planets, including Earth, orbiting the sun. He also concluded that we do not observe parallax because the stars are almost infinitely far away from the sun.

But Aristarchus was clearly ahead of his time. Most mathematicians and astronomers in ancient Greece considered the geocentric model of the universe to be a far simpler and more logical explanation of the movements of planets than the heliocentric universe of Aristarchus. Aristarchus was actually charged with impiety for removing the Earth and human beings from the center of the universe.

During several centuries of refining the Aristotelian model of the universe, Greek astronomers and mathematicians appeared to hold to the idea that the "heaviness" of the Earth and the motions of the spheres accounted for what we call "gravity"—the force that keeps the planets in their paths and the universe from dissipating. No one in those early times, according to the historical record, ever suggested that there was a universal force holding the stars and planets together and also governing the behavior of objects on the Earth. Aristotle's *Physics* contains no such suggestion and it was the accepted textbook for philosophers and scientists for almost 2,000 years. Up until the time of Galileo, scholarly work in the sciences consisted of adding corroborative details to Aristotle's great work—not attempting to replace it with other interpretations.

Meanwhile, as astronomical observations continued, the attempts to make Aristotle's model consistent with the data became quite contorted. One of the observational problems the Greeks encountered was that the planets did not seem to move in regular, spherical orbits in their supposed journeys around the Earth. The word "planet" comes from the Greek word *planete* or "wanderer," and wander is just what they appeared to do. We now know that the apparent zigzagging courses of the planets are simply due to the fact that we are observing them from the moving orbit of the Earth. Depending upon the positions of Earth and a planet we are observing, the planet may appear to be moving with us or in the opposite direction. The impious heliocentric model of Aristarchus would have dispensed with this puzzling problem immediately. But the popular geocentric model had to be forced to account for the planetary motions.

In the second century AD, the geocentric model peaked with Claudius Ptolemy (AD 83–161), whose elaborate Aristotelian concept was accepted for another 1,400 years, aside from some criticisms by Islamic scholars in the eleventh and twelfth centuries. Ptolemy's major work, *Almagest* (*The Greatest*), presents his model of the universe with the Earth at the center, surrounded by larger, transparent spheres containing the moon, sun, planets and stars, all following the accepted wisdom from Aristotle.

Where Ptolemy differed was that his planets actually moved in two circular orbits, a large one with the Earth as the focus, which he called the "deferent," and a smaller circle called the "epicycle" in which the planet traced a rosette pattern at the edge of the larger circle. Think of a ballerina

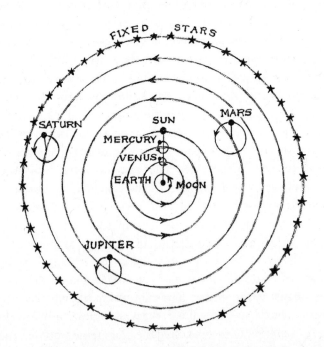

FIGURE 1: The Ptolemaic universe was the solar system with the Earth at its center.

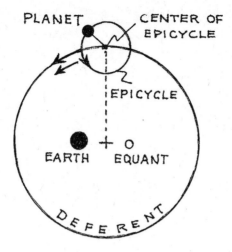

FIGURE 2: Ptolemy's idea of deferents and epicycles explained the wandering motions of the planets.

twirling around on a stage, tracing a large circle as she goes, or Disney's Mad Tea Party ride, in which the tea cups at the perimeter (filled with screaming children) spin around in their own small circles while at the same time the platform holding all the tea cups spins. Ptolemy explained that the rotation of the epicycles (the spinning tea cups) within the deferents caused the planets to move closer to and farther away from the Earth at different points in their orbits. Depending on their position in the epicycles, the planets would even appear to slow down, stop, and reverse direction. Straining further to accommodate astronomical data, Ptolemy introduced the concepts of "eccentric" and "equant" to explain more about the seemingly complex movements of the "wandering stars" and thus to bolster the prevailing paradigm.

Today Ptolemy's model seems absurd. It amazes us that intelligent philosophers and astronomers could have been so convinced by it. Yet the model was ingenious in that it did explain the strange motions of the planets. And to an intellectual tradition caught up in the beauty of circles, it may even have seemed attractive.

THE RENAISSANCE: CENTERING THE SUN

The Ptolemaic model of the universe was one of the longest running erroneous ideas in the history of science. It wasn't until the early sixteenth century that Nicolaus Copernicus (1473–1543) seriously questioned this geocentric model, and harked back to the forgotten ideas of Aristarchus.

Copernicus, a Polish canon and astronomer, studied Ptolemy's model

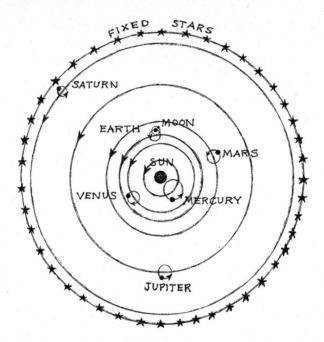

FIGURE 3: **The Copernican universe had the sun at its center but retained Ptolemy's deferents and epicycles.**

carefully, and found so many internal inconsistencies and clashes with his own astronomical data that he was driven to reconsider the basic assumption that all heavenly bodies orbited the Earth. He discovered that if the sun was placed at the center of the universe, then Ptolemy's system of epicycles could be greatly simplified. Although he retained Ptolemy's deferents and epicycles, and retained Aristotle's crystalline spheres, Copernicus did place the sun in its rightful place at the center of the solar system—or, as he would have said, at the center of the universe.

Copernicus spent most of his life working out the details of his new theory and refining his cosmology, but he wrote his book quietly, circulating his ideas mostly among friends, and did not allow it to be published until he was near death. As a canon in the Church, he realized that his revolutionary ideas would be viewed as heretical. Indeed, fifty-seven years after Copernicus's death, another forward thinking philosopher who embraced Copernicus's ideas, Giordano Bruno (1548–1600), was charged by the Inquisition and burned at the stake. In addition to his heretical ideas against Catholic dogma, Bruno maintained that the Earth moved around the sun, the sun moved through the cosmos, the universe was infinite, and the stars were suns with their own planetary systems.

One of the great doubters of Copernicus's sun-centered system was the most famous astronomer of his day, the Danish nobleman Tycho Brahe (1546–1601). His own ungainly geocentric model represents the last tweaking of the Ptolemaic system before the heliocentric system finally became widely accepted.

Brahe had a passion and a rare talent for astronomy almost from childhood, and made several important astronomical discoveries when he was quite young. He discovered that the conventional astronomical tables were inaccurate in certain important predictions. He observed and described a supernova, which changed in color from yellow to red. This was a shocking discovery because the "fixed stars" were supposed to be immutable. Tycho Brahe also made careful observations of a comet that appeared for ten weeks in 1577 and concluded that it was farther away from Earth than the moon and that its orbit was highly elliptical. Here were two more blows to the conventional cosmology that maintained that comets were located in the "sub-lunar" sphere close to the Earth, and that all heavenly bodies moved in perfect circles or spheres, not ellipses. Throughout his career, Brahe and his students catalogued the positions of 1,000 stars, overthrowing Ptolemy's catalogue, which was still the classic authority in the sixteenth century. All of these feats were achieved without the benefit of the telescope, which wasn't invented until almost a decade after Brahe's death.

Brahe had immense practical talent as an astronomer, but he was less talented mathematically and was unable to move cosmology forward to a truer model of the universe. He did dispense with Aristotle's crystalline spheres, but he confounded things further by proposing his own bizarre cosmology in which the Earth stood still at the center of the universe with two heavenly bodies orbiting around it: the moon and the sun. The sun, however, had the other five known planets orbiting around it at the same time that it orbited around the Earth!

In Brahe's time, there still was no scientific concept of gravity. But one reason that he was so attached to the idea of a geocentric universe was because of what he visualized as the heavy, unmoving Earth at the center, attracting all other bodies to it. Like many people who relied on common sense, he believed that if the Earth moved or rotated, a cannonball fired in the direction of rotation would go farther than one fired in the opposite direction, and since this was obviously untrue, he dismissed the notion of a moving Earth. Brahe did worry about the planets, once he had removed the spheres that had moved them. Having no concept of gravity, he wondered what would keep them from falling. But unwittingly, Brahe contributed greatly to the eventual realization of universal gravitation: His accurate measurements of the perturbations of the moon and the motions of planets would be used by

Kepler and Newton to arrive at the correct view of the solar system and the laws of gravity.

In 1596, Tycho Brahe read a curious little book about the solar system written by a young Austrian mathematics teacher, Johannes Kepler (1571–1630). The fanciful and mystical book explained the orbits of the planets around the sun, in the Copernican system, in terms of the five regular solids in geometry and the perfect circle. Tycho Brahe was impressed by the author's astronomical talents, because the mystical mathematical pattern accounted for the distances of the planets from the sun within only a 5 percent error. He invited Kepler to come to Prague and become his assistant. Tycho Brahe died only two years after their famous collaboration began, but Kepler remained in Prague and continued laboriously studying Tycho Brahe's voluminous astronomical data on his own. He had given up on his odd idea of the planetary orbits corresponding to the geometrical solids because the 5 percent error seemed too large to him. Eventually Kepler made a revolutionary discovery. Tycho Brahe had been most concerned about his data on Mars, but Kepler discovered that the increased accuracy of Tycho Brahe's data prevented him from fitting the motions of *any* of the planets into his Copernican model. However, he found that if he let the planets move in ellipses, with two foci, then everything seemed to work out. Kepler, a devout Lutheran who was just as attached to the Platonic notion of perfect, divine circles and spheres as the ancient Greeks had been, was nearly driven mad by his discovery.

As is often the case in science, once a dramatic leap forward is taken and all the facts click into place, then it seems obvious that what has been discovered is true. However, we must appreciate that at the time, Kepler's leap in thinking flew in the face of almost 2,000 years of study and consensus about the planetary system and the universe, and contradicted even Kepler's intellectual predecessor Copernicus and mentor Tycho Brahe by showing that nature does not behave in a perfect, symmetrical way. This is a marvelous example of the workings of science. Sometimes, to fit small observational facts into theory, we are forced to accept a huge change in our understanding of the universe. Kepler's new model of the solar system was the second time, after Copernicus, that such a dramatic paradigm shift was made based on compelling scientific data.

Kepler also took the heliocentric model a step closer to a modern concept of gravity. Kepler's sun, stripped of its crystalline spheres as Tycho Brahe's had been, exerted what Kepler called a "force" on the planets. He understood that it was this force that caused the differences in the speed of revolution of the planets: The farther away a planet is from the sun, the slower it moves, and the longer is its revolution or "year." Kepler got quite close to the concept of gravity, but he believed that the force binding the planets to the sun was some form of magnetism.

Through painstaking study and the courage to follow the data despite his own Christian and Platonic bent of mind, Kepler brought astronomy into the modern era. A central sun exerting a force on the planets sent the ancient crystalline spheres into oblivion. Kepler's realization that the planets moved in elliptical orbits similarly got rid of Ptolemy's (and Copernicus's and Brahe's) complicated deferents and epicycles. Kepler's famous three laws of planetary motion, published in the early seventeenth century, encapsulated the new and accurate mathematical treatment of the planetary system:

- Planets move around the sun not in circles but in ellipses, the sun being at one focus.
- A planet does not move with a uniform speed but moves such that a line drawn from the sun to the planet sweeps out equal areas of the ellipse in equal time.
- The square of the period of revolution of a planet round the sun is proportional to the cube of the average distance of the planet from the sun.

MOVING TOWARD GRAVITY

Galileo Galilei (1564–1642), born in Pisa, Italy, was a contemporary of Johannes Kepler. Although the two corresponded, Galileo never quite grasped the significance of Kepler's laws of planetary motion. He was convinced of the truth of the Copernican heliocentric universe, but was still enough of a traditionalist to reject the idea of imperfect elliptical planetary orbits. Still, Galileo is considered the father of experimental science. He was a remarkable man, combining craftsman skills with mathematical talents, and he was just as interested in the laws of motion on the Earth as he was in furthering knowledge in astronomy. He was also a gifted writer and a somewhat irascible individual. As is well known, the Church condemned Galileo's search for scientific truth; the Inquisitors arrested him near the end of his life and forced him to recant his belief in the sun-centered universe.

With the aid of a revolutionary new instrument invented by a Dutch spectacle maker, the telescope, Galileo made observations of the heavens that eventually overthrew Aristotle and Ptolemy and proved the truth of the heliocentric planetary system. Almost every significant thing that Galileo observed through his telescope contradicted the old, long-accepted Greek cosmology, even more dramatically than Tycho Brahe's observations had. The moon, Galileo saw, was not a perfect sphere as the Greeks had believed, but was pock-marked and jagged like the Earth. Not every celestial object revolved around the Earth, or even the sun: Galileo was astonished to find four moons orbiting Jupiter, a mini-Copernican system in itself. The sun had spots, another impossibility according to the ideas of the perfect universe, and their movement across the face of the sun indicated either that the sun moved or the Earth or both. When Galileo examined the Milky Way, he was amazed to see at least ten times more stars with his telescope than were visible to the naked eye. This observation seemed to corroborate the speculations of Giordano Bruno, burned at the stake less than a decade earlier.

Most important, Galileo was able to prove an observational prediction made by Copernicus more than fifty years previously. Copernicus had devised a test that could differentiate between his planetary system and Ptolemy's: observing the phases of Venus. Presciently, he believed that someday an aid to the naked eye would enable astronomers to see this phenomenon. According to Ptolemy's model, because Venus is circling the Earth and moving around in its own epicycle between Earth and the sun, observers on Earth would only see Venus in quarter-phase at all times. But in the sun-centered system, Venus would have phases like our moon, going from slice to quarter to half to full and back again. In a dramatic verification of the heliocentric model, Galileo observed and documented the moonlike phases of Earth's sister planet.

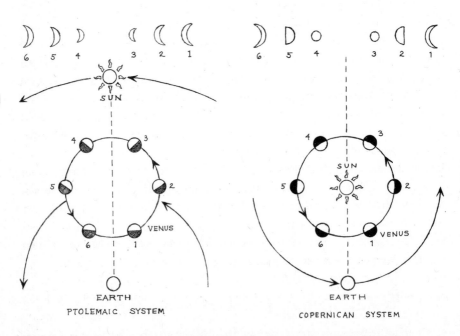

FIGURE 4: Galileo's observations of the phases of Venus confirmed the Copernican system.

Although Galileo is best known as an astronomer, his experiments on motion formed the foundation of the science of mechanics. In this sense, Galileo was the first physicist, and the first scientist to emphasize the importance of experiments in checking out theories. One of the most widely held beliefs in Renaissance science was Aristotle's doctrine of motion that stated, among other things, that the speed of falling bodies is proportional to their weight. This rule squared with common sense: Heavier objects would fall faster than lighter ones. But Galileo proved that this was untrue. It is probably apocryphal that he dropped cannon balls from the Leaning Tower of Pisa, but the experimental apparatus he built in his backyard allowed him to come to the same conclusions, and with more control over his experiments. Galileo built a series of inclined planes of different angles, and rolled heavy balls of different weights down them, timing their passage from start to finish.

Through his many experiments, Galileo discovered the "equivalence principle," which proved to be such a blow to the Aristotelian system: Bodies fall at the same rate (in a vacuum) independent of their composition. He also discovered the concept of inertia: A body continues to move with uniform speed unless stopped by a force or another object. He discovered that falling bodies accelerate at a constant rate. He determined that the distance traveled by objects falling for time t was proportional to their acceleration times the square of the time. He even determined that the acceleration was approximately 9.8 meters per second per second, and that the average velocity of an object that has been falling for time t is the acceleration times t divided by two.

These quantitative statements illustrate how Galileo brought mathematics and physics together; indeed, he turned physics into a science, and paved the way for Newton's even more monumental advances. Galileo even anticipated Einstein in showing that motion is relative. The example Galileo used was being inside a smoothly sailing ship—unless you look out the porthole, you do not sense that the ship is moving.

But did Galileo understand "gravity" and use the term? Although one of his early papers, published in 1589, concerned the center of gravity in solids, in his writings on motion he does not use the word gravity, but does refer to the "heaviness" of objects and the Earth. Galileo thought deeply about acceleration, but whether he considered "heaviness" to be related to, or even identical with, acceleration is not known.

THE ENLIGHTENMENT: "GRAVITY FROM NOW ON"

Fittingly, since he was in many senses Galileo's intellectual heir, Isaac Newton (1642–1727) was born in either the year that Galileo died or the year after, depending upon which calendar one consults. Like Galileo, Newton was just as interested in what happened in the solar system as he was in the motion of objects—and other phenomena such as color, optics, and tides—on the Earth, and he was just as convinced of the importance of the experimental method in advancing knowledge. Galileo had combined mathematics and physics, and Newton continued on this path, culminating in an astounding mathematical description of the universe and most of what physically occurs on the Earth, findings that he set out in his three-volume *Philosophiæ Naturalis Principia Mathematica,* published in 1687. Along the way, as a tool to help him express his ideas, Newton invented calculus.

In Newton's skillful hands, Galileo's "heaviness" became universal "gravity," the elemental force of attraction in the universe. Newton dazzled his scientific peers in England and Europe, including his enemies, of which he had several, and he was richly rewarded with honors during his lifetime for his achievements. Yet Newton was an odd personality: secretive, paranoid, and often ungenerous. When slighted or criticized, he would respond with vituperative anger, often in letters, or would simply retreat. Most of his scholarly work remained unpublished during his lifetime—millions of words on a surprising variety of subjects, including alchemy and religion.

When Newton was a student at Trinity College, he read Aristotle but also the writings of his near-contemporaries Galileo and the French philosopher-mathematician René Descartes (1596–1650). Descartes advanced the idea of deductive reasoning and stressed that mathematics was a general science that could be applied to all scientific disciplines. In this he was following Plato's idea that the route to new knowledge lay through mathematics. However, Descartes was less interested in testing mathematical ideas by confronting them with experiments. For example, he proposed the idea of "vortices" whirling in the "ether" to explain the motions of the moon and planets, a notion that would be very hard to test observationally. Newton, on the other hand, was very interested in the environment around him, particularly in the nature of matter, the motions of bodies, light, colors, and sensations. He was a natural experimenter, given to inductive reasoning. In his private notebooks at this time, Newton wrote extensively about the idea of atoms, the role of God in the creation of matter, infinite space, eternal time, designs for clocks, the nature of light, possible causes of the tides, and the nature of gravity.

A terrible plague hit England in the summer of 1665, and Cambridge University was closed for two years. During that time, Newton returned to his mother's home in Woolsthorpe, Lincolnshire. This provided him with even more time to think and work on his own. He concentrated on his mathematical studies, inventing new ways to compute logarithms and finding a way to calculate the area under any curve known to mathematicians at the time. This is when he invented the calculus, or as he called it, the "fluxional method," to be able to describe more accurately flow and rates of change. He explored the concept of the "infinitesimal," counterpart to the infinite universe at the other end of the distance scale. And, as he related later in life, he was inspired by comparing a falling apple in the Woolsthorpe orchard with the moon to wonder whether they were attracted to the Earth by the same force. "I began to think of gravity extending to the orb of the Moon . . . ," Newton wrote, "and computed the force requisite to keep the Moon in her Orb with the force of gravity at the surface of the earth . . ."[1]

Newton had asked himself the question: Why does the moon not fall down onto the Earth the way an apple does? At first he considered whether it was because the moon was not subject to the same force of "heaviness" that drew objects to the ground. But then he understood that, theoretically, if a rock or cannonball is thrown forward with enough speed, it will overcome the pull of the Earth and go into orbit; the rock would then never hit the ground. Newton's next realization was that the attractive force of gravity is not unique to the Earth, but must act between all bodies in space. This was a remarkably revolutionary idea and eventually brought about another paradigm shift. It directly contradicted the Aristotelian belief that the laws governing the motion of bodies in the heavens were completely different from (and more pure than) the laws of physics here on Earth (in the corruptible sublunar world). Newton realized that he could explain all the motions of the planets in the solar system, as well as the moon's, using one universal law of gravity.

While at Woolsthorpe, Newton studied Kepler's writings, which reinforced his understanding that the moon's rotation around the Earth is associated with a centripetal force that exactly balances the attractive pull of Earth's gravity. Given the mass of the moon and its velocity, it could only be located at the precise distance from the Earth where indeed it is found. Newton then wrote his insights down in mathematical terms: The force of gravity is proportional to the masses of the two bodies attracting one another and is inversely proportional to the square of the distance between them. Fitting the numerical values for the moon's orbit into his equation, Newton came up with a proportionality constant for his force law, which could be applied anyplace in the universe where there is a gravitational force. This number is now known as Newton's gravitational constant. Using his new equation and constant, Newton predicted the period of the moon as approximately twenty-eight days, close to the known period of 27.322 days. (Twenty years later, while writing Book Three of the *Principia*, Newton recalculated the moon's period to be twenty-seven days, seven hours and forty-three minutes—rounded off to 27.322 days.)

Discovering the law of universal gravity was not a one-time *Eureka!* event for Newton. He worked on the ideas that had come to him in his early twenties for many years, and did not begin publishing them in his *Principia* until he was in his mid-forties.

Newton's lifelong rival Robert Hooke had reached similar ideas about gravity independently, yet he was unable to express them mathematically. Both Newton and Hooke agreed that there was an attractive force pulling objects toward the center of the Earth. Both had rejected Descartes's ideas of vortices and ether. Both knew, due to the work of Kepler and his laws for planetary motion, that an inverse square law may cause a body to orbit in an ellipse. But Hooke was unable to work out the mathematics for planetary ellipses, and knowing of Newton's legendary mathematical abilities, in 1680 he wrote to Newton proposing that Newton determine the mathematical ellipse corresponding to the motion of a planet and provide the physical reason for it. Hooke even stated: "My supposition is that the Attraction always is in a duplicate proportion to the Distance from the Center Reciprocall"[2]—that is, it varies in accordance with the square of the distance between two bodies. Newton did not respond to this letter.

Four years later, in August 1684, when Newton was forty-one, his friend Edmond Halley, the discoverer of the comet later named after him, posed Newton a very similar question: If we are given an inverse square law of attraction to the sun, what would the orbit of a planet look like? In fact, as has often been the case in science, a significant advance was "in the air." Hooke, Halley, and the architect and astronomer Christopher Wren all claimed to have had the idea of an inverse square law for gravity by thinking about Kepler's third law, although they were unable to express it mathematically. Everyone now knew from Kepler that planets move in ellipses, but the ellipse had yet to be connected to a law of gravity. Newton knew the answer to Halley's question, since he had calculated it almost twenty years previously. He devoted some months to finding proof of the statement that a planet moves in an ellipse according to the inverse square law of gravity. At this time, there was no precise terminology for these basic ideas in mathematics and physics, so Newton invented them. For example, he called a quantity of matter "mass;" the quantity of "motion" Newton defined as the product of velocity and mass. He sent the initial results of his work, *On the Motion of Bodies in Orbit*, a slim, nine-page treatise to Halley, who begged to publish it, as his one copy was so much in demand.

Newton was suddenly on fire. He wrote to the Astronomer Royal and others, asking for more data: on comets, stars, the moons of Jupiter, a table of tides. He virtually locked himself in his room at Trinity, ate and slept erratically, and worked feverishly for months, producing thousands of pages of manuscript, until he had finished the first volume of his masterpiece, one of the greatest works in science of all time. In the *Principia*, Newton laid down three laws of motion:

1. Every body perseveres in its state of being at rest or of moving uniformly straight forward, except insofar as it is compelled to change its state by forces impressed. (This is the law of inertia.)
2. A change in motion is proportional to the motive force impressed and takes place along the straight line in which that force is impressed.
3. To any action there is always an opposite and equal reaction; in other words, the actions of two bodies upon each other are always equal and always opposite in direction.

The second law translates into the mathematical statement that the force on a moving body is equal to the mass of the body times its acceleration. For the gravitational force between bodies, Newton incorporated the inverse square law: The force is proportional to the inverse square of the distance between the two bodies.

Newton worked on successive editions of the three-volume *Principia* for much of the rest of his life. By the time he finished the final volume, he had shown that gravity is the cosmic glue that holds the solar system together ("All the planets are heavy toward one another") as well as ruling motion on the Earth. "It is now established that this force is gravity," Newton wrote, "and therefore we shall call it gravity from now on."[3]

It took 2,000 years to get from Aristotle to Newton, to graduate from a naïve, stationary Earth–centered view of the universe to the brilliant comprehension of a universal force of gravity holding the heavenly bodies in orbit as well as explaining the behavior of objects on the Earth. Newton's universal law of gravity has remained a major pillar of science for more than 300 years. Yet Newton did not consider his great work on gravity to be complete. He was troubled by not fully understanding the motions of the moon, and often talked about wanting to solve this problem, even in his old age. Also, he was greatly disturbed by not being able to identify the "cause" of gravity. In Newton's day, especially among those who could not relate to his mathematics (and they were numerous), the idea of an invisible force acting at a distance on objects smacked of the occult. Newton was well acquainted with the occult and the metaphysical, having written far more, though unpublished, words during his long life on alchemy, religion, and mysticism than on science. Yet he never allowed these research topics to impinge upon his mathematical science. We can almost hear the regret in the words with which Newton admitted defeat: ". . . we have explained the phenomena of the heavens and of our sea by the power of gravity, but have not yet assigned the cause of this power . . . I have not been able to discover the cause of those properties of gravity from phenomena, and I frame no hypotheses . . ."[4] Newton left a deeper explanation of gravity to future generations of scientists.

When I was a student at Trinity College, Cambridge, in the 1950s, Sir Isaac Newton's presence was still almost palpable. Often, when walking along the path across the Great Court leading to Trinity's Great Gate, I would ponder the law of gravity that Newton had discovered. Despite the fact that Newton's theory of gravity was modified by Einstein, it still remains valid today, supporting and challenging physicists after all this time. I sometimes paused outside the rooms that had been Newton's when he had performed his remarkable optics experiments and when he had written his awe-inspiring *Principia*. I remember entering the Trinity College antechapel and standing in front of Roubiliac's life-sized marble statue of Newton shining in the half-light, and being struck with wonder at the ingenuity of the human mind and the miracle of scientific discovery.

Chapter 2

EINSTEIN

I saac Newton was Albert Einstein's hero. "Newton, forgive me;" Einstein wrote in 1932, by which time his new theory of gravity, general relativity, was well established and accepted by most of his peers, "you found the only way which in your age was just about possible for a man with the highest powers of thought and creativity. The concepts you created are guiding our thinking in physics even today, although we now know that they will have to be replaced by others further removed from the sphere of immediate experience, if we aim at a profounder understanding of relationships."[1]

Although Einstein built his theories from facts and sought to verify theories with data, he distrusted "common sense" and the evidence of the senses—the "sphere of immediate experience," as he expressed it in his musings to Newton. Einstein entertained counterintuitive notions that allowed him to pull physics from the mechanistic, clockwork universe of the eighteenth century up into the twentieth century. The mental puzzles—such as chasing a light beam and never catching it—led Einstein to his new understanding of the universe and the laws of physics.

Comparing Newton's and Einstein's concepts of gravity is like comparing Ptolemy's Earth-centered universe to Copernicus's sun-centered one: Both seem to fit the available data almost equally well—for example, both seem almost equally borne out by astronomical observations of the planetary orbits—but the pictures they present of "reality" are profoundly different. In Newton's theory, and in his cultural worldview, space and time were separate and absolute. That is, space was the absolute arena of simultaneous locations of material bodies, and time was the absolute measure of the motion of bodies in space. In Einstein's theory, on the other hand, space and time are not different entities, but form a four-dimensional "spacetime." The laws of physics must appear the same to all uniformly moving observers. In Einstein's spacetime, whether two events are simultaneous is a relative concept, depending on the motion of the observer.

Newton's gravity is a force between bodies, and it works very well to explain the motion of planets in the solar system. Einstein transformed the very idea of gravity: from Newton's universal force to a new understanding of gravity as geometry, as the warping, or distortion, of spacetime by the presence of matter or energy. This notion that gravity results from the geometry of spacetime can explain in a natural way why you cannot tell the difference between gravity and acceleration. In his gravity theory, Newton equated the gravitational mass of a body, responsible for producing the gravitational force, to the inertial mass of a body, without a fundamental understanding of the origins of this phenomenon. In Einstein's theory, on the other hand, the geodesic motion of a body in curved spacetime naturally incorporates the equivalence of inertial and gravitational mass.

Newton's and Einstein's concepts of gravity are profoundly different. Yet deciding by experiment, observation, and data which one is more correct involves surprisingly small numbers and tiny effects. Only 43 seconds of arc per century distinguish between Einstein's calculations of the precession of Mercury's orbit and those obtained using Newton's gravitational theory. Einstein's prediction of the redshift of light emitted by atoms in a gravitational field had to wait for several decades before technology was advanced enough to be able to even detect this small effect. The famous prediction of the bending of light by the sun during a solar eclipse had to wait for forty or more years to be fully confirmed by accurate measurements of light emitted by distant quasi-stellar objects.

Yet the wonder is that Einstein's new theories, both special relativity and general relativity, proved to be testable by observation and experiment within his lifetime. More than ninety years later, physicists are still working out solutions to Einstein's equations to uncover new consequences. The public as well as many scientists are still struggling to grasp the meaning of his work.

RELATIVITY IN THE AIR

A radical change in the conception of space and time was in the air in the early twentieth century. Physicists began questioning the absoluteness of space and time that was an integral part of Newtonian mechanics. In the nineteenth century, Michael Faraday and James Clerk Maxwell had made major discoveries about electromagnetic phenomena. In particular, the experimental verification by Heinrich Hertz of Maxwell's prediction that electromagnetic waves move at the speed of light had a profound influence on the development of physics in the early twentieth century. Guglielmo Marconi's success in transmitting radio waves across the Atlantic in 1901 also motivated physicists such as Einstein to rethink the fundamental understanding of the significance of the finite speed of light.

Centuries before, Galileo had understood the notion of the relative motion of inertial frames of reference in a limited way, pertaining to material bodies moving much more slowly than light.* The fastest forms of transportation in the sixteenth and seventeenth centuries were ships or horse-drawn carriages, at orders of magnitude smaller than the speed of light. To go beyond Galileo's limited slow-motion relativity principle, Einstein had to imagine what would happen if bodies moved close to or at the speed of light. Einstein had been thinking about the electromagnetic properties of light for many years. As early as age sixteen, he had pondered what would happen if you chased a light beam. He concluded that you could never catch up with the light beam, and this insight already prepared the way for a radical alteration of our understanding of space and time. Thus the ideas of special relativity that burst from him onto paper in 1905 had been incubating for years.

When Maxwell made his remarkable discovery of the unified field equations of electricity and magnetism, published in 1861, he firmly believed that the electromagnetic waves had to move through a medium that could support the wave motion, and this medium was called "the ether." In later years, Maxwell even promoted a mechanical model of the ether, involving gears and cog wheels. Like Ptolemy's epicycles, the concept of the ether was one of the longest-running erroneous ideas in the history of science. It had prevailed since the time of Aristotle, who postulated besides the elements of earth, fire, air, and water, a fifth element called "quintessence," in which were embedded the crystalline spheres that supported the circling planets and the sun. As water is to fish, so was the ether to the planets, including the Earth and all that lived on it. To the Greeks, bodies in motion had to be in contact with *something* in order to move, and that something was a colorless, invisible medium. Most scientists in the nineteenth and early twentieth centuries also believed, with Maxwell, that the ether was something more than a vacuum, something less than air, through which the stars and planets moved, and through which electromagnetic waves were propagated. By the late nineteenth century, it was almost universally accepted that the ether

---

* An inertial frame of reference is the vantage point of an observer who moves with constant speed, without any external force exerting an influence.

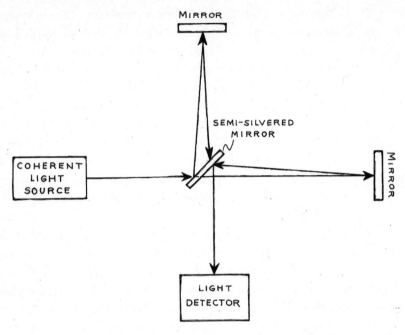

FIGURE 5: Michelson and Morley's interferometry experiment did not detect the ether.

existed everywhere in space and that light traveled through that medium even in "empty" space. Light traveled as electromagnetic waves, and the speed of light was measured with respect to the supposed ether. But the ether had yet to be detected, and some doubted its existence.

In 1887 at the Case Institute of Technology in Cleveland, Albert Michelson and Edward Morley designed a crucial experiment to resolve the question of the ether. They built an interferometer, which consisted of a light source, a half-silvered glass plate, two mirrors on separate arms, and a telescope, which functioned as a light detector.

The experiment would allow a measurement of the speed of light in different directions, thereby measuring the speed of the ether relative to Earth, and establishing its existence. Michelson and Morley were attempting to detect shifts of interference fringes between light waves that had passed through the two perpendicular arms of the apparatus. Such shifts would occur, it was believed, because the light would travel faster along an arm oriented in the same direction as the moving ether and slower when the arm was oriented in a perpendicular or opposite direction. If the two arms were positioned differently with respect to the supposed motion of the ether, then the crests and the troughs of the light waves would arrive at the end point with different synchronization, shifting the fringe pattern. The equipment was accurate enough to detect a difference of only one ten-billionth in the speed of light going in the direction of the motion of the Earth and in the direction perpendicular to this motion.

However, the experiment did not detect any difference in the speed of light, no matter which direction it traveled in, or how the interferometer arms were oriented, or where the Earth was in its annual orbit around the sun. Instead, both light rays arrived back at their starting points at the same time, in unison, with no shift in interference patterns, no matter how many times or under what conditions Michelson and Morley repeated the experiment. Their reluctant conclusion was that the ether did not exist.

Yet twenty years later, in Einstein's day, physicists were still arguing about the ether. Despite the null results of the famous experiment, most physicists were loathe to give up the time-honored ether. Both Hendrik Lorentz and the Irish physicist George FitzGerald, for example, attempted to retain the concept of the ether in their analyses of the experiment. Lorentz concluded that either the ether simply did not exist, or that there was some strange, unknown characteristic of matter or space and time that had not been discovered yet but would explain the odd behavior of the ether.

He suggested that electrical charges created by the ether moving against an object would disturb the particles in the object and contract it in the direction of movement, thus explaining the null result of the Michelson-Morley experiment. FitzGerald also proposed the contraction model to explain the null result. He suggested that all moving objects would be foreshortened in the direction of movement—and hence the instruments themselves in the Michelson-Morley experiment must have been shortened in relation to the light ray moving with and against the ether, so that the apparatus could never truly detect the ether. The measuring device he used as an example was a foot ruler: A moving ruler would actually be very slightly shorter than a ruler at rest.

The French mathematician Henri Poincaré was also pursuing relativity. "Suppose that in one night all the dimensions of the universe became a thousand times larger," he wrote. "The world will remain similar to itself . . . Have we any right, therefore, to say that we know the distance between two points?"[2] That is, the concepts of space and distance are entirely relative to the frame of reference that one is in. In 1904, a year before the publication of Einstein's famous papers, Poincaré gave a speech at the World's Fair in St. Louis in which he suggested that a "whole new mechanics" was necessary in order to pull physics out of the crisis it had fallen into after the Michelson-

Morley experiment. In this new mechanics, he continued, the speed of light "would become an impassable limit." The paper Poincaré presented in St. Louis described a relativity theory based on inertial frames of reference, in which, unlike in Newton's theory, there was no absolute meaning to space and time. He was not at first satisfied with the FitzGerald-Lorentz hypothesis, yet like the two physicists, Poincaré was unable to free himself entirely from the concept of the ether.

EINSTEIN'S SPECIAL RELATIVITY

The findings of Michelson and Morley and the ideas of FitzGerald, Lorentz, and Poincaré led right into the concepts of relativity. What Einstein did was to dispense with the ether entirely, sharpen arguments for the relativity of inertial frames, and come up with the dramatic results showing that matter and energy are equivalent and that the speed of light is a constant, representing the limit of velocity.

"On the electrodynamics of moving bodies," published in June 1905, outlines the special theory of relativity. In this paper Einstein did away with the ether and made the speed of light an absolute constant. Like FitzGerald and Lorentz, Einstein predicted that in special relativity, rods are contracted in their direction of motion when moving close to the speed of light. Moreover, he stated, clocks slow down when moving with a speed near that of light. Einstein independently derived the transformation law that kept Maxwell's electromagnetic field equations invariant when transforming from one inertial frame of reference to another (that had been published a year earlier by Lorentz and are now called the "Lorentz transformations"). These transformations from one uniformly moving frame of reference to another form the basic mathematical equations underlying special relativity. Once Einstein postulated these transformations together with the absolute and constant speed of light, then all the consequences for the physics of spacetime in special relativity followed in logical order. In particular, Einstein was able to derive the generalization of Newtonian mechanics and the correct equations of motion for material particles, which conformed to the laws of special relativity. The famous thirty-page paper contains no references, although Einstein thanked his friend Michele Angelo Besso for helpful discussions. In particular, there were no references to the earlier papers by Michelson and Morley, Poincaré, and Lorentz, which Einstein claimed in later years not to have known about.

Einstein's paper entitled "Does the inertia of a moving body depend upon its energy?" of September 1905 is only three pages long. It contains the most celebrated equation in physics—$E = mc^2$—telling us that the energy of a body equals its mass times the square of the speed of light. Since the speed of light is so huge (186,000 miles, or 300,000 kilometers, per second), when squared and multiplied by even a small mass, the energy produced is enormous. Einstein did not foresee the practical nature of his discovery at the time he published his paper. Other scientists understood the possibility of using nuclear fusion and fission for military purposes and this eventually gave birth to the atomic bomb. Much later, during the Second World War, Einstein wrote a letter to President Roosevelt urging him to consider making an atomic bomb before the Nazis were able to produce one. Under Julius Robert Oppenheimer's leadership, the Manhattan Project made the bombs that were dropped on Hiroshima and Nagasaki in August 1945.[3]

Few people would dispute that special relativity is Einstein's theory, despite the contributing research from some of his peers. Einstein's main paper of 1905 was a clear and logical presentation of the new concepts of space and time, released from the irrelevant notion of an ether. Outside of the mainstream of physics at the Swiss patent office in Bern, outside of the academic environment, perhaps his mind was more free to roam: to ignore the problematical ether, to tie the speed of light securely to an absolute value, to conceive of the unity of matter and energy. Special relativity, a wonder in itself, was also in Einstein's mind a stepping stone. The special case would lead to the general. For the next ten years, Einstein struggled to develop his masterpiece, general relativity, which would incorporate gravity.

GENERALIZING RELATIVITY

From 1907 onward, Einstein thought about incorporating gravitation into relativity theory. He was concerned about two major limitations of the special theory of relativity. The first was that it was restricted to inertial frames of reference. What happens if you describe the laws of physics in an accelerating frame of reference? The second was that special relativity did not include Newton's law of gravity in a self-consistent way. Newton's theory of gravity was based on the concept of action at a distance, that is, the force of gravity between two bodies acted instantaneously. This could only happen if the information about the gravitational force was transmitted between the two bodies with an infinite speed. This clearly violated the first principle of special relativity, that no body or information could move faster than the speed of light. In short, gravity had to be compatible with special relativity. In order to accomplish this compatibility, special relativity had to be generalized to include gravitation.

In 1922, in a lecture Einstein gave when touring Japan with his second wife, his cousin Elsa, he recalled the moment when he had had the germ of the idea of general relativity.* One day while working at the patent office, Einstein related to his audience, he almost fell out of his chair when a powerful thought struck him: If a man jumped off a high building, he would not be able to tell whether he was falling due to the pull of gravity or whether some force was exerting an acceleration upon him. In other words, you cannot distinguish between gravity and acceleration.

Thinking about the man falling off the building led Einstein into thinking about Galileo's experiments with falling bodies and the principle of equivalence: Bodies fall in a gravitational field independently of their composition. For Einstein the equivalence principle meant that gravity and acceleration were the same. That is, the inertial mass of a body is equal to its gravitational mass; the inertial mass of a body, which resists a force, is equal to the body's mass that causes gravitational acceleration. While Newton had simply assumed this to be true in his gravitational theory, Einstein made the principle of equivalence a cornerstone of his developing theory of gravitation. The difference between Newton and Einstein was that Einstein recognized the principle of equivalence as a fundamental, motivating principle in gravity.

Recall that Newton was unhappy with two aspects of his theory of gravitation: the instantaneous or "action at a distance" phenomenon associated with gravity, and the fact that he could not say what the "cause" of gravity was. Einstein was able to clear up the first aspect of this confusion fairly quickly, for according to special relativity nothing moves faster than light, not even gravity. Newton's instantaneous gravity was obviously impossible. As for the second, transforming gravity from a force to the geometrical curvature of spacetime did away with the problem: The "cause" of gravity was matter warping spacetime.

Getting from the falling man to the final version of the theory of general relativity was a long and arduous intellectual effort, taking Einstein almost a decade, with serious stumbling blocks along the way. He had only one collaborator at two stages of the journey, his friend Marcel Grossmann.† At times he worked as if in a fever, knowing he was in a race with other talented physicists and mathematicians.

Indeed, unlike in the development of special relativity, Einstein was not a lone genius working out the ideas of the new gravity theory outside of the

---

* It was on this trip, on the ship bound for Japan, that Einstein learned he had won the 1921 Nobel Prize for physics for his 1905 discovery of the photoelectric effect.
† Einstein's friend Besso also helped with important calculations such as the calculation of the precession of Mercury's orbit.

mainstream of physics. Several other well-known physicists and mathematicians were also attempting to describe gravity as a generalization of special relativity. The aims of all these scientists, including Einstein, were to determine how fast gravity traveled and what in fact gravity *was*, and when those issues were settled, to find a gravity theory that incorporated the finite—as opposed to the Newtonian instantaneous and therefore infinite—speed of propagation of the gravitational force.

As early as 1900, for example, at a meeting of the Amsterdam Academy of Sciences, Hendrik Lorentz presented a paper entitled "Considerations of gravitation." Einstein greatly admired Lorentz, considering the older physicist a charming person and a mentor. At the meeting, Lorentz noted the fact that so many phenomena had been successfully accounted for, by applying electromagnetic theory that it seemed natural to him that gravity should also be propagated at the speed of light. This outright statement was in contrast to the opinion of an earlier, prominent investigator, the French mathematician Pierre-Simon Laplace, who had claimed that gravity should propagate with a speed vastly greater than light.

There were some strange ideas circulating as scientists struggled to understand gravity better. Back in 1758, Georges-Louis Le Sage had proposed that "corpuscles of particles" could produce pressures on a body, thus causing the force of gravity. In this proposal, Le Sage anticipated the modern idea of the "graviton," the particle that may carry the force of gravity just as the photon carries the electromagnetic force. In the nineteenth century, Le Sage's theory became popular when it was studied by William Thomson (Lord Kelvin) in the context of the newly discovered kinetic theory of gases. However, Maxwell in 1875 and Poincaré in 1908 helped to discredit it.

At the 1900 Amsterdam meeting, Lorentz sought to explain gravity in terms of electric and magnetic interactions. He considered the possibility that Le Sage's "ultra-mundane corpuscles" could be replaced by electric waves that could exert a pressure against a body to produce gravity. However, further considerations led him to conclude that this hypothesis must be wrong.

An even stranger idea of gravity arose from the "one-fluid" concept of electricity, which was arrived at independently in the mid-eighteenth century by the famous American scientist and statesman Benjamin Franklin, the English physician and scientist William Watson, and the German Franz Aepinus. It was further developed in 1836 by Ottaviano-Fabrizio Mossotti. The idea was that electricity is "one fluid" made up of atoms that repel each other. Atoms in solid materials were also supposed to repel each other. But the atoms of the ether were slightly attractive to these atoms, creating a net attraction. According to this theory, the combination of these forces constituted gravity.

Poincaré, in 1904, was the first to point out that if relativity theory was true, then gravity must propagate with the speed of light. He proposed his own modification of Newtonian gravity, which was further developed by Hermann Minkowski, a Swiss mathematical physicist, in 1908 and Willem de Sitter in 1911. Poincaré, Minkowski, and de Sitter agreed that the modification of Newton's gravity theory would require corrections to the perihelia of the planets. However, the correction they came up with for Mercury, which has the largest perihelion advance of all the planets, was not big enough to account for Mercury's actual anomalous motion.

In Berlin, the celebrated Max Planck became involved with issues of gravity in 1907. He began by considering the work of the Hungarian Count Roland von Eötvös, who had established experimentally in 1891 that the inertial mass of a body and its gravitational mass are always exactly equal. Planck then said that all energy has inertial properties and must gravitate. Six months after Planck's paper on the subject, Einstein introduced his principle of equivalence, showing that gravitation equals acceleration.

Although several prominent physicists described gravity theories that were consistent with special relativity in that the speed of the propagation of gravity was the speed of light, none of these theories was satisfactory. These special relativistic theories were limited to the physics of inertial frames of reference. Einstein himself created one such unsuccessful theory before arriving at general relativity, and he realized its limitations.

BEGINNING THE JOURNEY TO GENERAL RELATIVITY

Einstein's first attempts to generalize the special theory of relativity were based on a theory of the scalar field that obeyed an equation saying that gravity was transmitted at the speed of light. The idea of a "field"* originated with the research of Faraday and Maxwell, who used this concept to explain the forces of electricity and magnetism as they were transmitted with the finite speed of light between static or moving electrical charges.† The scalar

---

* The notion of a field is demonstrated when metal filings form a "field pattern" on a piece of paper suspended above a magnet.
† A scalar field theory of gravity was also developed by the Finnish physicist Gunnar Nordström, who likewise attempted to modify Newtonian gravity to make it consistent with special relativity and the finite speed of light.

nature of the field reflects the fact that in contrast to a "vector" field, it has no direction in space. It is a comparatively simple mathematical scheme, involving only one component of a field in space and time. The gravitational field satisfied an equation that was invariant under the Lorentz transformations of special relativity.

In special relativity the only admissible reference frames are those moving relative to one another with uniform speed, whereas in reality reference frames can move at different nonuniform speeds and in different directions. Think of the different perspectives of passengers of two airplanes flying in different directions, the drivers of cars below, watching them, pedestrians watching the cars and planes, and an astronaut in space watching the Earth. The three axes of space (x, y, and z) determine a spatial position, while the time obtained from reading a clock determines the time an event occurs within a certain reference frame. Minkowski formulated the mathematical description of Einstein's special relativity as a four-dimensional spacetime—three spatial and one time dimension—that played a significant role in the future development of relativity theory.*

Will arbitrarily moving observers have different experiences of reality? In the years 1912 to 1914, Einstein had the brilliant insight that the laws of physics must be independent of the frame of reference in which they occur. This meant that whatever frame of reference a physicist uses to make physical measurements of phenomena, even an accelerated one, the laws of physics remain the same. Einstein called this universal invariance of the laws of physics "the principle of general covariance." It motivated in large part the mathematical formulation of his new gravity theory, and became a fundamental cornerstone of general relativity.

The notion of an absolute space had troubled Newton. This absolute space distinguished from all others one uniformly moving, nonrotating inertial frame of reference at rest, to which we refer all physical laws. In particular, in the Newtonian world, the absolute space determined which events occurred simultaneously with respect to the absolute frame of reference. In special relativity, on the other hand, the simultaneity of two events is a relative concept, depending upon which reference frame one is in. Newton claimed that even a single body in an otherwise empty universe possessed inertia, that is, the tendency for a body to continue moving with constant speed or

---

* Minkowski was Einstein's former mathematics teacher at the Eidgenössische Technische Hochschule (ETH) in Zurich. He had considered Einstein a lazy student, and had been less than helpful promoting him to academic positions.

to remain at rest in the absence of an external force. This claim had played a crucial role in his favoring an absolute space. Newton considered inertia to be the primary property of a body in absolute space, while gravitation was a universal but secondary property.

In 1689, Newton demonstrated his idea of absolute space with his famous bucket experiment, which we can easily replicate. Fill a bucket half full with water and tie a rope to the handle. Twist the rope repeatedly, rotating the bucket until you cannot twist the rope any longer. Now steady the bucket, let the water settle, and let go of the bucket, allowing it to spin by the rope. At first, the water inside does not rotate with the bucket and its surface is flat. But eventually the water begins to rotate with the bucket and its surface begins to appear concave. Finally, as the rope begins to twist in the opposite direction, the rotation of the bucket slows down and the water is now rotating faster than the bucket and its surface remains concave.

Newton asked the question: "Why does the surface of the water appear concave?" The answer would seem to be that the rotation of the water makes the surface concave. But at that point when the water's surface is concave, it is actually *not* rotating relative to the bucket. When the bucket is first released and starts rotating, the water rotates relative to the bucket and its surface is flat. Only when the friction between the water and the sides of the bucket causes the two to rotate in unison, with no relative motion between them, does the water surface become concave. We infer that the shape of the water's surface is not determined by the rotation of the water relative to the bucket.

Newton then thought about performing the bucket experiment in empty space. In an empty universe, what can it mean to have the bucket and water rotating? There is nothing to refer the rotation to. How could we even know the bucket was rotating? Newton decided there had to be something against which one could measure the bucket's rotation, and *that something had to be space itself*. When the water is not rotating with respect to absolute space, then its surface is flat. On the other hand, when the water rotates with respect to absolute space, it appears concave. In Book One of the *Principia* Newton wrote: "Absolute space by its own nature, without reference to anything external, always remains similar and unmovable." Yet Newton was dissatisfied with this situation, for he continued: "It is indeed a matter of great difficulty to discover and effectually to distinguish the true motions of particular bodies from the apparent, because the parts of that immovable space in which these motions are performed do by no means come under observations of our senses."

It would be many years before Newton's idea of absolute space was contested. The English philosopher George Berkeley claimed that the water became concave not because it was rotating with respect to absolute space, but because it was rotating with respect to the fixed stars, or all the rest of the matter in the universe, as reference points. The German philosopher and physicist Ernst Mach carried Berkeley's argument further, seriously challenging Newton. In 1883 he wrote,

> Newton's experiment with the rotating vessel of water simply informs us, that the relative rotation of the water with respect to the sides of the vessel produces no noticeable centrifugal forces, but that such forces *are* produced by its relative motion with respect to the mass of the Earth and the other celestial bodies. No one is competent to say how the experiment would turn out if the sides of the vessel increased in thickness and mass until they were several leagues thick.[4]

Mach argued that if all the matter in the universe were absent, then the surface of the water in the bucket would never become concave. Obviously we cannot perform this experiment with an empty universe, so we do not know whether Newton or Mach was right.

Mach demanded that the entire system of mechanics should be reformulated in terms of a concept of motion of material bodies *relative* to one another. He proposed that inertial frames of reference should be determined on the basis of relative motions of bodies in the universe. Mach clearly showed the limits of Newtonian classical mechanics. He explicitly related the concept of the inertial frame to the motion of all the rest of the mass in the universe. That is, the mass of the universe determined the inertial mass of any given body. The property of the inertia of a body is still a controversial issue among physicists, and speculations about its cause are still proposed in scientific papers.

Einstein perceived the inertial property of a body and its gravitational property in a different way from Newton. He perceived them to be the same, and emerging from the geometry of spacetime. This equality could not be correctly explained within a purely special relativistic theory of gravity. Einstein had to get rid of the basic asymmetry between inertial and gravitational mass in Newtonian mechanics, in which the inertial mass is a property of an isolated body in an otherwise empty universe. Gravitational mass was conceived by Newton as a property of a system of bodies. For Mach, the inertial mass of a body is determined by the mutual accelerations within a given system of bodies, not as an independent property of a single body. Strongly influenced by Mach, Einstein introduced the principle of equivalence, which expressed the *equality* of inertial and gravitational mass independently of the specific laws of motion in classical mechanics.

In 1913, Einstein wrote a letter to Mach in which he supported Mach's interpretation of Newton's bucket experiment and said that it agreed with his own ideas of a new theory of gravitation. Einstein even included Mach's principle in his general relativity in the form of the equivalence principle, which has been tested with great accuracy on Earth. Following Mach, Einstein stated that the behavior of the water in Newton's rotating bucket is determined by the gravitational forces of all the matter in the universe.[5] Throughout the years in which Einstein developed his general theory of relativity, he remained strongly influenced by Mach's philosophy.

THE GEOMETRY OF SPACETIME

In the summer of 1912, Einstein had the insight that non-Euclidean geometry should play an important role in his new theory of gravitation. He realized that there was a problem with special relativity, a problem that later became known as Ehrenfest's paradox. The circumference of a uniformly rotating disk should shorten due to Lorentz contraction as measured by an observer at rest. On the other hand, the radius of the disk should remain the same, for it is perpendicular to the motion of the disk. Under these circumstances, the ratio between the circumference of the disk and its radius is no longer given by 2π, as in Euclidean geometry. Einstein deduced from this that Euclidean geometry is no longer valid in a theory generalized to include rotating frames. This was a significant intellectual breakthrough, and paved the way for his general theory of relativity.

The deeper Einstein went into his theory, the more dissatisfied he became with the mathematics available to him to describe his thoughts. His friend Marcel Grossmann, a talented mathematician who became dean of the mathematics department at the ETH in Zurich where the two had originally met as students, suggested that he consider using the non-Euclidean geometry of the nineteenth-century German mathematician George Bernhard Riemann.

Riemann was a brilliant mathematician many years ahead of his time. The famous geometry that he developed describes curved surfaces on which parallel lines can intersect, which is impossible in Euclidean geometry. Riemann conceived of the idea that matter would warp geometry, although he applied this insight to a three-dimensional non-Euclidean geometry; the idea of spacetime in four dimensions was still more than fifty years away. Riemann also concluded that geometry can cause force, which at that time was a momentous insight.

Einstein took up Grossmann's suggestion, and incorporated Riemannian geometry into his developing theory, but he did not stop there. He also adopted the notion of a "metric tensor," which had been developed by Riemann's professor, Carl Friedrich Gauss, as well as the differential calculus of the Italian mathematician Tullio Levi-Civita. The metric tensor determines the infinitesimal distance between two points in spacetime. It produces the now familiar grid that makes up the curving geometry of spacetime.[6] The sum of the angles of a triangle on a sphere is not equal to 180 degrees as it is on a flat Euclidean surface. A metric tensor is the mathematical description of the geometry of a surface, for example, a flat plane and a sphere.

Without the tool of the metric tensor, it would not be possible to describe a curved, four-dimensional spacetime mathematically. According to Einstein's thinking, the metric tensor determines not only the strength of the gravitational field but also the distance scales in every direction in space and time, as well as the rate of clocks in the gravitational field. The metric tensor consists of ten functions of space and time, and all ten of these degrees of freedom—except for four so-called gauge degrees of freedom—contribute to the description of the curving gravitational field.

Another crucial tool in forging general relativity was the mathematical formalism invented by Minkowski, which Einstein had already used as the natural mathematical language with which to describe the spacetime for special relativity. Combining Riemannian geometry with the formalism of tensor calculus, Einstein was able to develop field equations for gravity and nongravitational physics, such as electromagnetism, that were compatible with his principle of general covariance and the principle of equivalence.

One of the basic features of Riemannian geometry is the so-called Riemann curvature tensor, which determines the amount by which four-dimensional spacetime is warped. When this curvature tensor is equal to zero, the metric tensor becomes that of Minkowski's flat Euclidean spacetime for special relativity, and this signals the complete absence of a gravitational field. However, in general relativity, the correct interpretation of special relativity is the observer's ability to neglect gravity, or the curvature in a local patch of spacetime. In this way, special relativity is included naturally in the geometrical formalism of general relativity.

OBSTACLES TO FINISHING THE THEORY

At this point in developing the details of the new gravity theory, Einstein ran into serious difficulties. In 1913 he and Grossmann published a paper in two parts, one by Einstein and one by Grossmann. In Grossmann's part of the paper, he proposed that a test particle would move along the straightest path in spacetime, which is known as a geodesic. Einstein had to develop, in his part of the joint paper, a set of field equations that would describe the geometry of spacetime (gravity) in the presence of a massive body. This set of equations had to reduce to the basic differential equations of Newtonian gravity, namely the Poisson equations invented by the French mathematician Siméon-Denis Poisson.[7]

Three related problems that Einstein and Grossmann ran into in this joint paper were serious. One problem was that they had violated the principle of causality. In technical terms this says that macroscopically an event in spacetime must happen only in the future light cone of another event. In other words, you cannot see an "effect" before its "cause" has occurred. (The term "light cone" simply denotes the cone formed in four-dimensional Minkowski spacetime by the rays of light arriving at or emanating from an event.) The light cone separates spacetime into a past cone and a future cone which meet at the point in spacetime corresponding to the event.

The second problem was that Einstein's original field equations violated the conservation of energy! On one side of the field equations, he had quantities that described the warping of the spacetime geometry. On the other side of the field equations, he had the energy momentum tensor that warped the spacetime geometry. The problem was that the geometrical left-hand side of the equation did not properly obey the condition leading to the conservation of the matter and energy on the right-hand side of the equation. A basic mathematical property of the geometrical side of the equation was missing. In other words, Einstein had not yet discovered the correct law of gravity. Needless to say, violating the sacred law of conservation of energy in physics was unacceptable.

Thirdly, for weak gravitational fields and slow-moving bodies, Einstein's equation did not properly reduce to Poisson's equation for Newtonian gravity. This was obviously a serious problem because it was well established that Newtonian gravity agreed with the phenomena of falling bodies on Earth and the motion of planets in the solar system, the one exception being the anomaly of the perihelion advance of Mercury.

The problems in the 1913 paper were obvious to other physicists. At Einstein's inauguration as professor in Berlin in 1914, Planck described Einstein's current research on gravity. He chided Einstein for having recently published a paper with Grossmann proving that you cannot construct a generally covariant theory of gravitation in the manner that Einstein had proposed.

For a while, Einstein was discouraged about his pursuit of a generalized gravity theory. However, with his characteristic tenacity and independence of thought, Einstein continued his gravity research in Berlin. By 1915, he was working strenuously to overcome the difficulties of 1913 and to break through to the last stage in the development of his gravity theory. The biggest problem, he realized, was that in his formulation of the gravitational field equations, he was missing something in the description of the curvature of spacetime.[8]

SPACETIME BANISHES VULCAN

In early 1915, in spite of the problems he faced—and perhaps to get his mind off of these problems for a while—Einstein turned to the well-known topic of the anomalous perihelion advance of Mercury. He wondered whether his new gravity theory might explain this critical anomaly without recourse to a dark, still unseen planet, even though his theory was not yet complete. So Einstein solved his field equations for the weak gravitational field determining Mercury's orbit using perturbation theory, because he assumed that his corrections to Newtonian gravity would be small.*

Fortunately, the missing part of Einstein's field equations did not have to play a role in this calculation. The sun, treated as the source of gravity in the solar system, was considered a point source, so that Einstein was able to use his gravitational field equations that were valid in empty space. Remarkably, the solution of the field equations of empty space gave the correct answer for the anomalous advance of the perihelion of Mercury, even though the equations did not yet contain the extra piece crucial to guarantee the conservation of energy in the presence of matter.

Einstein's calculations revealed that he could account for the tiny difference between Newton's prediction of Mercury's perihelion advance and the actual observed advance by plugging just the mass of the sun and Newton's gravitational constant into his equations. It was not necessary to take into account the mass of Mercury, because in comparison to the huge sun, it could be treated as a test particle with vanishingly small mass. Fitting the anomalous Mercury data with his new, partially completed gravity theory was an astounding result and a high moment for Einstein, because he now knew that he was on the right track. His theory was able to explain precisely what had been an astronomical mystery for decades. "For a few days I was beside

---

* Perturbation theory is a mathematical approximation that assumes that solutions of differential equations can be obtained by adding up quantities that in each order are smaller than the preceeding ones, and that the final sum of these quantities generates the solution of the differential equations.

myself with joyous excitement,"[9] Einstein wrote to one friend, and admitted to another that the discovery had given him heart palpitations.

This is a marvelous example of how physics should be done: The deep mathematics of a new theory confronts a long-standing observational anomaly and explains it without recourse to any free, adjustable parameters or "fudge factors." Thus with the publication of the Mercury paper in 1915, even though Einstein had not yet completed general relativity, and there were still serious problems to solve, the scientific world moved a step closer to accepting the idea that gravity was actually geometry and the curvature of spacetime rather than Newton's instantaneous dynamical force.

GRAVITY AS GEOMETRY:
FINISHING GENERAL RELATIVITY

The last six months of 1915 were a critical time for Einstein in developing general relativity. Europe was already one year into the Great War. Einstein, thirty-five years old then, was working feverishly on his equations while a great slaughter was taking place in the muddy trenches of France and Belgium, and many of his young German countrymen were dying. While the armies clashed, Einstein focused his attention on the final steps in completing his monumental gravity theory. He had to find the final physically consistent form of his gravitational field equations in the presence of matter. This was the missing formulation of his theory that he had been searching for.

The famous German mathematician David Hilbert had heard about Einstein's revolutionary—though as yet unfinished—work on gravity and invited him to give a series of lectures at the University of Göttingen in 1915. Einstein accepted the invitation and gave seven hours of lectures over a period of a week, introducing the mathematicians and physicists at Göttingen to his new ideas.

David Hilbert was one of the world's most renowned mathematicians and an expert on Lagrangian and Hamiltonian methods in theoretical physics. The French mathematician Joseph Louis Lagrange had developed a method to determine Newton's laws of motion and celestial mechanics that was based on the principle of least action discovered by another French mathematician, Pierre-Louis Moreau Maupertuis, who had published his celebrated principle in 1746. The famous Irish mathematician, Sir William Rowan Hamilton, had made important contributions to the dynamical theory of Lagrange and the variational calculus. Thus Hilbert was in a strong position to help Einstein solve the last remaining problem in the formulation of general relativity: the final form of the basic field equations that would describe matter and spacetime geometry, avoid the fatal violation of the conservation of energy, and correctly yield Newtonian gravity for weak gravitational fields.

Soon after Einstein's departure from Göttingen, Hilbert, excited by the possibilities Einstein had presented during the previous week, sat down and derived the correct field equations for Einstein's gravitational theory using a world function called a "Lagrangian function" or, in a transformed version, a Hamiltonian function. These functions were invariant under general coordinate transformations and yielded the correct gravitational field equations.[10]

Hilbert submitted his paper on this work to the *Göttingen Nachrichten* in November, 1915 and sent a postcard to Einstein in Berlin, enthusiastically explaining his results. This communication lit a fire under Einstein, prompting him to derive the correct field equations independently by brute force. After considerable effort—because he was not as gifted a mathematician as Hilbert—he obtained the same result.

There has been sharp debate in recent years over who was responsible for the correct derivation of the field equations of general relativity—Einstein or Hilbert?—but this disagreement misses the point, because the fundamental ideas and the structure of the theory were entirely Einstein's. Although the final derivation of the field equations was almost a trivial matter for a mathematician with Hilbert's skills, no one would dispute that general relativity was Einstein's theory.

The notion of the curvature of spacetime being determined by the presence of matter became the fundamental basis of Einstein's gravity theory. This represented a huge leap forward in our understanding of gravity and indeed the structure of the universe. It was an important departure of thinking, in that when compared to other forces of nature such as electromagnetism, gravity was now the geometry of spacetime and the very arena in which the other forces played out their roles. Einstein now proceeded to finish his gravitational theory according to the rules of tensor calculus.

Einstein's final exposition of his great work was "The foundation of the general theory of relativity," published in *Annalen der Physik* in 1916. In this paper, he incorporated Maxwell's theory of electromagnetism in a way that made Maxwell's equations invariant under any general coordinate transformations. Thus he succeeded in describing the two fundamental forces known at this time—electromagnetism and gravitation—within one elegant geometrical framework. He included in his paper the correct answer for the anomalous advance of Mercury's perihelion (which had been published the previous year), the gravitational redshift formula, and the predicted result for the bending of light grazing the limb of the sun.

When Karl Schwarzschild published later in 1916 an exact solution of

Einstein's field equations for spacetime, describing a static, spherically symmetric gravitational field with a single massive particle at the center, Einstein's gravity theory achieved a fairly complete form. Schwarzschild worked out the solution of Einstein's field equations for a nonrotating, spherically symmetric body—which became the basis of black holes—while serving on the eastern front in World War I. He died in May 1916, at age forty-two, shortly after the publication of this important solution, of a rare autoimmune disease he developed during the war.

TESTING AND CONFIRMING THE NEW THEORY

Back in 1911, working only with Newton's theory, Einstein combined Doppler's principle with his principle of equivalence. The Doppler principle (or "effect" or "shift") is named after its discoverer, the Austrian physicist Christian Doppler, who in the mid-1800s first described how the observed frequency of light and sound waves is affected by the relative motion of the source and the detector. What Einstein proposed was that a spectral line emitted by an atom situated in a place of strong gravity—for example, on the surface of the sun—would be shifted slightly toward the red, longer-wavelength end of the spectrum. This would happen because in a strong gravitational field, a distant observer would see atoms vibrating more slowly than in a weaker gravitational field like the Earth. This is the celebrated redshift—or "Einstein shift"—prediction that had to wait for confirmation until technology and astronomy had advanced sufficiently. Einstein followed up this work with a paper in *Annalen der Physik*, also in 1911, that contained another idea for experimental verification. He argued that since light is a form of electromagnetic energy, it must gravitate. Thus a remote observer would see that a ray of light passing near the strong gravitational field of a body such as the sun must be curved—and the velocity of light must depend on the gravitational field.

Einstein had the right idea in 1911, but his calculations of the bending of light had been off by a factor of two. It is fortunate that in 1914 when he asked his friend, the astronomer Erwin Freundlich, whether he could attempt an observation of the bending of light by the sun during a solar eclipse, Freundlich was unable to carry out the project due to the onset of the war. If Freundlich had been able to do the experiment, Einstein would have been shown to be wrong in his prediction, because he had used Newton's gravity theory to predict the bending of light, not his own evolving theory. In 1915 Einstein calculated, using the geodesic equation for a light ray, the amount of bending that a light ray from a distant star would experience as it "grazed the limb of the sun"—or passed close by the sun on its journey to observers on Earth. Einstein again used perturbation theory, as he had done with his calculations for Mercury, and found that the light ray should be bent by an angle of 1.75 arc seconds toward the sun.

It is easy to get caught up in the excitement of the science during these dramatic years of paradigm shifting and forget the context in which it took place. Between 1914 and 1918, Germany and Britain, along with France, Belgium, Russia, Turkey, and, as of 1917, America were at war. Despite his Swiss passport and ideals for world government, Einstein was widely considered a German scientist. Not only had he been born in Germany, but he spent all the war years living and working in Berlin. This context makes the years 1916 through 1919 all the more remarkable for the development of general relativity, for it was predominantly English, Dutch, and Belgian astronomers who interpreted and verified general relativity, thereby pushing it and Einstein into worldwide public prominence.

In 1916, as the battles raged on the western front in Belgium and France, Einstein sent a copy of his final paper on general relativity to his Dutch colleague, the astronomer Willem de Sitter. De Sitter, who was associated with the British Royal Astronomical Society, in turn sent the paper to Arthur Eddington, the secretary of the Society. Eddington was not only a brilliant astronomer but also a fine mathematician, and he immediately understood the importance of Einstein's work. At thirty-four, just two years younger than Einstein, Eddington was climbing the academic ladder rapidly, and was eager for a new challenge.

From the very beginning, there was excitement among British astronomers at the possibility of testing Einstein's theory during a solar eclipse. There were several solar eclipses coming over the next few years, and the one that was judged the best for observations would be on May 29, 1919. And so, in the midst of the war, with German U-boats tightening their noose around Great Britain, Eddington and other British astronomers began planning the expeditions that three years later would verify the theory of a "German" physicist, making him within a few short months the most famous scientist in the world.

In the spring of 1919, after the 1918 armistice, Eddington and his colleagues set off for the island of Principe in West Africa, while two other British astronomers headed for northern Brazil, to capture on photographic plates the May 29 solar eclipse as it moved across the Atlantic. There were high expectations for good results that would either prove or disprove Einstein's theory, because the eclipse would be taking place against the background of a group of very bright stars, the Hyades in the constellation Taurus. Despite some disappointing clouds in West Africa, both groups took photographs, and analysis began even before returning home to England.

But the British astronomers were tight-lipped for several months. An impatient Einstein wrote to his friend Paul Ehrenfest in Leiden in September 1919, asking if he had heard anything about the expeditions, and Ehrenfest passed on Einstein's query to Lorentz. At the end of September, after some sleuthing, Lorentz sent a telegram to Einstein telling him the results, which were not made public until a joint meeting of the British Royal Society and Royal Astronomical Society on November 6, 1919. On that day, under the gaze of Sir Isaac Newton, the former president of the Royal Society whose portrait hung in the meeting room, the astronomers formally announced their results. Those working in Brazil had obtained a value for the bending of light of 1".98 ± 0."30, and Eddington's group at Principe had found it to be 1".61 ± 0."30, both agreeing within their margins of error with Einstein's prediction of a deviation of 1".75. The chairman of the meeting, Joseph John Thomson, declared, "This result is one of the greatest achievements of human thought."[11]

During the years after the Royal Society announcement, there were debates about the actual size of the errors involved in the analysis of the photographic plates obtained by Eddington's team. The weather conditions during the eclipse observed in West Africa were not perfect and the analysis techniques were far from satisfactory. Due to the cloudy weather, Eddington was only able to obtain sixteen photographic plates, and of these, only six contained stars sufficiently close to the limb of the sun that they could be identified as sources whose light could be bent. He had to average the measurements, and by a judicious choice of plates ended up with an averaged measurement that was consistent with Einstein's prediction.

Although Einstein always maintained that he *knew* the solar eclipse data would verify his gravity theory, his excitement—if not producing the heart palpitations of the earlier Mercury results—was evident in the postcard he sent to his dying mother, Pauline: "Good news today. H. A. Lorentz has wired me that the British expeditions have actually proved the light shift near the sun."[12]

In fact, in the history of general relativity, the experimental confirmation of the bending of light was a more momentous event than the fitting of the anomalous Mercury data. In the case of the perihelion advance of Mercury, Einstein had shown that his theory fitted and explained a well-known anomalous piece of data. It was not necessary to go hunting for new data. But in the case of the bending of light, Einstein made a prediction based on his theory, and astronomers went out to collect the data that could have either confirmed

Einstein's prediction or disproved it. This is why the scientific community created such an event in announcing the results, and why the press turned it into news that flew around the world. The fitting of the Mercury data had not received nearly so much attention.

However, acceptance of the bending of light confirmation and Einstein's new gravity theory was not universal. Old paradigms die hard. This new and mathematically difficult theory would be replacing Newton's. There was, for a time, significant opposition to that idea. In particular, famous American astronomers such as William Campbell and Heber Curtis of the Lick Observatory in California did not readily accept Einstein's theory or the English bending of light measurements as conclusive evidence for the correctness of the theory. The Lick Observatory's ongoing evaluation of the 1918 Goldendale eclipse in the United States had produced a negative result for the bending of light, although Campbell was not willing to go public on this result.

At the same time, astronomers such as Campbell, Charles Edward St. John, and the English astronomer John Evershed in India were attempting to measure the gravitational redshift of the frequency of light from the sun, which was Einstein's third test of general relativity. These measurements were complicated because of the need to model both the sun's corona and distorting effects in the sun's interior. The lack of success in obtaining a definitive result for the gravitational redshift consistent with Einstein's prediction led to further doubts about the validity of the new gravity theory.

An American naval astronomer, Captain Thomas Jefferson See, published a book in which he claimed that the perihelion advance of Mercury could be calculated from Newton's gravity theory without relativity, and he also cast doubts on the results of the 1918 and 1919 solar eclipse data. Even some English astronomers and physicists, such as Ludwig Silberstein, opposed Einstein's theory and published papers claiming that the theory had serious theoretical deficiencies compared to Newtonian gravity. In addition, the German astronomer Hugo von Seeliger stated publicly that the mathematics of Einstein's general relativity were too difficult to understand.

After the 1919 solar eclipse expeditions and the Royal Society announcement, Einstein had to face severe opposition not only from astronomers but from the physics community. One anti-Semitic faction of German physicists mounted a public opposition to Einstein's gravity theory. They rented a concert hall in Berlin for two days to present talks explaining why Einstein had to be wrong, and why his work should be ignored. Einstein responded with a letter to the editor of a Berlin newspaper, hotly defending his gravity theory and showing disdain for his critics.

But after 1922, most of the opposition to general relativity began to die down. In that year, solar eclipse observations in Australia and on Christmas Island in the Pacific vindicated the 1919 Eddington results. William Campbell himself, from the Lick Observatory, led the astronomy team in Australia, where perfect weather allowed successful observations. Campbell had developed new camera techniques and methods to analyze the eclipse observations. After months of careful analysis of the data, the Lick team, in collaboration with the Swiss astronomer Robert Trumpler, obtained a result for the bending of light that was consistent within the errors with Einstein's prediction of 1.75 arc seconds.

The observations of the Lick team did sway much of the physics and astronomy community to accept Einstein's gravity theory as being correct. Yet when Einstein was awarded the Nobel Prize, it was not for his work on relativity. The Nobel committee stated that the prize was being awarded for Einstein's work on the photoelectric effect and other contributions to physics in 1905.

Nevertheless, Einstein's general relativity was recognized eventually to be a significant paradigm shift. It would have enormous implications for the future of understanding the universe at both small astronomical distance scales as well as the large-scale structure of the universe.

PART 2

THE STANDARD MODEL
OF GRAVITY

Chapter 3

THE BEGINNINGS OF MODERN COSMOLOGY

Gravity plays a dominant role in our understanding of the large-scale structure of the universe, because it is the only force that operates significantly over the huge distances of the cosmos. The development of modern cosmology engaged the ideas and skills of some of the greatest minds of twentieth-century physics and astronomy on both the theoretical and observational sides. Beginning in the 1930s and continuing today, the accumulation of accurate data by powerful telescopes and space missions has turned cosmology into a science. Increasingly, we are able to test theoretical ideas about the structure of the universe and its beginnings. This modern science had its origins with Einstein and the pioneers of twentieth-century astronomy.

THE STATIC VERSUS DYNAMIC UNIVERSE

Einstein invented the modern science of cosmology in 1917 with his paper "Cosmological considerations on the general theory of relativity." Here he considered for the first time how the large-scale structure of the universe could be pictured in his gravitational theory. In order to make his dynamical field equations consistent with his idea of a static, unchanging, eternal universe, Einstein introduced a mathematical term that he called the "cosmological constant" into those equations. By adjusting the magnitude of this constant to yield the hypothesized finite size of the universe, Einstein's vision of the universe became compatible with the generally accepted static model that began with the Greeks. Einstein clearly had qualms about introducing the cosmological constant. In his paper, he wrote: "First of all we will indicate a method which does not in itself claim to be taken seriously; it merely serves as a foil for what is to follow."

Why was Einstein so convinced that the universe did not have a beginning, that it was always and forever the same? Part of the reason was his characteristic personality bent toward harmony and unity. Philosophically and perhaps spiritually, Einstein resisted the idea of a violent, sudden beginning to matter, energy, and spacetime. Then too, the consensus of his peers in the early twentieth century was that the universe was static and unchanging, and for once, Einstein went along with the scientific herd. Later regretting this lapse, he called the cosmological constant his "biggest blunder." It was a fudge factor to make his equations produce the results he wanted them to.

It took someone else perhaps less constrained by philosophy, or removed from the mainstream of European physics, to work out Einstein's field equations in their pure form without recourse to the cosmological constant. This was the Russian cosmologist and mathematician Alexander Friedmann. He published two papers in 1922 and 1924 showing that Einstein's field equations had three possible solutions, depending upon what he called the "curvature constant" of spacetime. Friedmann's model was of an expanding universe: When his curvature constant was zero, then the universe was spatially flat and infinite, but when it had a positive or a negative value, the universe would be spatially closed like a balloon, or spatially open like a saddle, respectively. In all three of these possible configurations, the universe according to Friedmann's solutions of Einstein's equations was not static, but was a dynamic expanding or contracting universe.* In addition to his physics papers, Friedmann published a popular book titled *The World as Space and Time*, introducing the idea that the universe had a beginning far in the past— perhaps "tens of billions of years" ago, that it appeared from "nothing," and that it might be oscillating through expansions and contractions.

Einstein objected to Friedmann's conclusions so much that he published a letter in *Zeitschrift für Physik* in 1922, claiming that Friedmann had made a mathematical mistake. Eventually Einstein had to admit that he had been in error and that Friedmann's calculations were correct. However, he still objected to the physical consequences of the solutions: that the universe had a beginning and that it was constantly changing. Tragically, Friedmann died

---

* The idea that a non-Euclidean geometry such as spacetime could have these three possible descriptions was first understood by the Russian mathematician Nikolai Lobachevski, and independently by the Hungarian mathematician Janos Bolyai, back in the early nineteenth century, although they did not think of applying their mathematical ideas to the universe.

in 1925 at the age of thirty-seven from typhoid fever. He never received credit for his remarkable discoveries about the universe, and died virtually unknown to the scientific community at that time.

Although the static model of the universe appeared to continue without further opposition, the dynamic model did not die with Friedmann. It was independently discovered by a Belgian, Georges-Henri Lemaître, who had an unusual double career as a Roman Catholic priest and astronomer. In 1927, Lemaître proposed the idea of an expanding universe in an obscure Belgian journal.

Einstein was not any more positive about Lemaître's ideas than he had been about Friedmann's. When the two met for the first time in 1927, Einstein surprised Lemaître with the news that Alexander Friedmann had already come up with the idea of an expanding universe, and added bluntly that although he couldn't fault Lemaître's mathematics, his physics was "abominable." Stinging from Einstein's criticism, Lemaître nevertheless went on to develop what seemed to him the natural conclusion of his theory: An expanding universe must have begun with a violent explosion, starting from a relatively small, dense object that he called the "primeval atom" or the "cosmic egg." Lemaître was also aware of the radioactive elements and their decay into other chemical elements, and he proposed that the elements that we observe today were due to the decay of the original primeval atom.

The cosmology community stood with Einstein, staunchly resisting Lemaître's exploding cosmic egg idea until astronomical data confirmed it in 1929. Partly it was a concern about religion. Although Lemaître tried hard to keep his religion and his science separate, he sometimes ran into trouble when others assumed his science was corrupted by a Christian bias, for a beginning to the universe implied a creator. Lemaître served as an initiator of the debate over whether and how the universe began, but refused to become actively involved in it. Much later, in 1951, Pope Pius XII published an unusual statement that amounted to an endorsement of Lemaître's model by the Catholic Church. According to the pope, a moment of creation of the universe proved the existence of God. Lemaître was upset by this papal intrusion into the realm of science, and personally convinced Pius XII to desist from further attempts to relate the developments of modern cosmology to Catholicism.

HOW BIG IS THE UNIVERSE?

The 1920s was a fertile decade in the history of astronomy. Modern cosmology was being born. There was lively debate over the two models of the universe, the static and dynamic views. There was also a related debate about the size and extent of the universe—specifically about whether the smudges of light known as nebulae, discovered in the eighteenth century by the French astronomer Charles Messier, were objects within our Milky Way galaxy or whether they were located far beyond it. That is, was the Milky Way the entire universe or just a small part of it?

The eighteenth-century German philosopher Immanuel Kant had conjectured that Messier's nebulae were distant "island universes" outside our Milky Way galaxy, but many scientists in the early twentieth century disagreed. A "Great Debate," sponsored by the National Academy of Sciences in Washington, DC, took place in 1920 between the astronomers Harlow Shapley and Heber Curtis, from California's Mt. Wilson and Lick observatories, respectively, to attempt to resolve this issue. Curtis strongly believed that the nebulae were distant collections of stars outside the Milky Way, whereas Shapley stuck to the idea that the nebulae existed only within our universe, that is, the Milky Way galaxy. Although by most accounts the debate resulted in a draw, Shapley was in effect rewarded for his defense of the establishment position with the directorship of the prestigious Harvard College Observatory.

There were new tools and concepts in the 1920s to help resolve the debate over the extent of the universe. This was the beginning of the era of the giant optical telescopes, when observational cosmology took great leaps forward. Since the invention of the daguerreotype and photographic plates, precise observations of stars and galaxies could be made at leisure, away from the telescope. Photography also made it possible to observe the atomic spectral lines emitted by stars, which led to exciting new knowledge about the composition and movement of stars. Such knowledge would not have been possible without similar leaps forward in theory. In the early 1900s, the Danish physicist Niels Bohr and the New Zealand physicist Ernest Rutherford were developing a model of the atom. Bohr introduced the idea that the electrons orbiting the nucleus of an atom radiated energy in quantum "lumps," in contradiction to the classical notion that radiation energy was continuous. These new theoretical insights fed into the young science of cosmology, as astronomers grappled with how the chemical elements we see today could have evolved from an early universe of hydrogen.

To decide whether the nebulae were inside or outside of our galaxy, one would obviously need to have a method of measuring stellar distances. How far away were the stars? How far across was the Milky Way? How far away were those blurry nebulae? Even measuring distances within the solar system had been a challenge that took many centuries to solve. It took until the late seventeenth century for astronomers to figure out reasonably accurately the mean distance between the Earth and the sun, a measurement that came to be known as one astronomical unit. Gauging the distances to the stars was much more difficult. Only a few stars are close enough to us that we can observe their motions against the background of farther "fixed stars" and calculate their distances from us through parallax and trigonometry. By the early 1900s, the distances to almost 100 stars had been triangulated, by a clever method of using the Earth's position at opposite sides of its elliptical orbit around the sun (positions we occupy six months apart) as the base of the imaginary triangle reaching out to the nearest stars. By this method astronomers discovered, for example, that Alpha Centauri, the nearest star, is twenty-five trillion miles or 4.3 light years away.

When one looks up at the sky on a clear, moonless night in the country, the sprinkling of stars across the blackness is almost dizzying. Are the bright stars closer than the faint ones? Not necessarily. Stars have "absolute" and "apparent" brightness. Absolute brightness is their intrinsic brightness— that is, how bright they actually are, if you could see them from the vantage point of a spaceship only a light year or so away. Their apparent brightness is how bright they appear to us, which is a combination of their absolute brightness and their distance from our solar system.

Over 2,000 years ago, the Greek astronomer Hipparchus devised a way of categorizing stars according to their brightness. He labeled the brightest stars "first magnitude" and the faintest stars visible to the eye "sixth magnitude," with the remaining stars falling on the scale in between. In this scheme, the larger the magnitude number, the fainter is the star. When the telescope was invented in the early seventeenth century, this magnitude scale had to be changed because telescopes revealed stars that had simply been invisible before. Today the largest telescopes reveal stars that exceed twenty-third magnitude. The brightest objects in the sky actually have negative magnitude values. So for example, the brightness magnitude of the sun is –26.7, whereas the star Sirius has a magnitude of –1.4 and Arcturus, –0.04. This magnitude scale, of course, only refers to the apparent brightness of the stars, and tells us nothing about their distance from us, or their true, absolute brightness.

How do we determine distance? How do we tell the difference between a faint star fairly close to us and a very bright star that is far away? Fortunately, the universe contains some curious, blinking stars that were discovered in the early twentieth century and were able to act as cosmic yardsticks. A variable star varies in the intensity of the light it emits, and this variation can be either irregular or periodic. To determine the nature of a star's variability, one plots the star's apparent brightness against time, usually from days up to several weeks. The resulting curve shows the maxima and minima of brightness and reveals whether there is any periodicity, that is, whether the curve repeats itself in a regular fashion. About 30,000 known variable stars have been observed, and they are grouped into two broad classes: 1) pulsating and erupting variables, in which the variation in light intensity is an intrinsic feature caused by physical changes within the star, and 2) eclipsing variables, in which the cause of the variability is external—with two or more stars rotating around and eclipsing each other in our line of sight.

A rare class of variable stars of the first type is the Cepheid variables. These are yellow supergiant stars that pulse in a regular pattern due to periodic heating up and cooling down. One of the most famous Cepheid stars is the North Star, or Polaris, which is forty-six times larger than the sun. In the early twentieth century, it varied 8 percent from its average magnitude over a 3.97-day period. Since the middle of the twentieth century, however, Polaris has been varying less and also getting brighter. It appears to be 2.5 times brighter today than when Ptolemy classified it as a third magnitude star in antiquity, which has implications for the speed of stellar evolution.

In the early part of the twentieth century, a quiet, unassuming, deaf astronomer named Henrietta Swann Leavitt turned Cepheid variables into the long-sought cosmic yardstick that made it possible to measure distances to stars and other galaxies. This tool ultimately resolved the debate about the size of the universe. In 1902, when Leavitt began working at the Harvard College Observatory, she was assigned a project on variable stars. She scrutinized plates of the southern sky taken at the university's observatory in Peru, comparing plates taken of the same area of the sky at different times, hunting for variable stars.

Leavitt made her great breakthrough in understanding the nature of the periodicity of the Cepheid stars by concentrating her efforts on the Small Magellanic Cloud and the Large Magellanic Cloud, named after the explorer Ferdinand Magellan, who had used these starry "clouds" for navigation when sailing in the southern hemisphere. Leavitt identified dozens of new Cepheid variables within the Magellanic Clouds. She reasoned that the stars in these relatively small regions would be approximately the same distance away from the Earth. Therefore if a Cepheid in the cloud stood out in brightness from the others, its absolute or inherent brightness must be greater. Leavitt then plotted the varying brightness, or magnitude M, of many of the Cepheids against time and discovered a dramatic correlation: The larger the luminosity or magnitude of the Cepheid, the longer the period between maxima in brightness. In papers published in 1908 and 1912, she showed that this rule would work for Cepheids anywhere in the universe. Thus Leavitt's amazing contribution to astronomy was to provide a relative yardstick for distances in the universe: Astronomers now understood that Cepheid variables with the same period were of the same intrinsic brightness, no matter what their apparent brightness might be to observers. If two Cepheids are discovered in different parts of the sky with similar periods, and one is four times fainter than the other, then according to the inverse square law for the transmission of light, it must be twice as far away.

Others beginning with the Danish astronomer Ejnar Hertzsprung then used the parallax method to determine the actual distances to the closest Cepheid variables. Hertzsprung did this in a very clever way. The great eighteenth-century astronomer William Herschel had discovered that the sun and its solar system are moving through space in the spiral arm of the Milky Way. Later astronomers clocked the speed of this movement at about twelve miles per second or thirty million miles each year. Hertzsprung used the solar system's movement to create a platform for triangulating distances to Cepheid variables within the Milky Way. Then, using Leavitt's relationship between period and absolute brightness, he calculated Leavitt's Small Magellanic Cloud to be 30,000 light years away. Thus astronomers had calibrated Leavitt's relative yardstick into an absolute measure. Now, providing there was a Cepheid variable in the vicinity, they were able to determine the true distance of any object in the sky.

HUBBLE AND THE MODERN TELESCOPE

The person who finally resolved both astronomical debates—about the size of the universe and whether it was static or dynamic—was the American astronomer Edwin Powell Hubble. In 1919, after serving in the postwar occupation forces in Germany, Hubble went to Mt. Wilson Observatory in Pasadena, California, where he began observing the skies with the most powerful telescope in the world, the recently completed 2.5 meter/100-inch Hooker Telescope. Indeed, the rapid development of American optical telescopes in the early twentieth century is a fascinating topic in itself, involving competition between millionaires whose sights were trained on the heavens. Without these instruments, whose powers would have utterly astonished Galileo, the great astronomical debates about the size, age, and origins of the universe could not have been resolved.

Before the war, when Hubble was working on his PhD at the Yerkes Observatory near Chicago, he began a careful study of Messier's nebulae, and attempted to classify them as either intra- or extragalactic varieties, thus stepping early on right into the controversy over the size of the universe. The Smaller and Larger Magellanic Clouds, Hubble felt, were good candidates for possible extragalactic structures, and he took a serious interest in them. He was inspired by Leavitt's work on the Cepheid variables, which implied that the universe did not begin and end with our galaxy. While working on the Yerkes 40-inch/1-meter telescope, Hubble was well aware that much more information could be obtained using larger telescopes with greater light-gathering power and better resolution.

When Hubble arrived at Mt. Wilson in 1919, in one respect he was odd man out. Most of the other astronomers there favored the view that the Milky Way represented the entire universe and there were no significant structures outside of it. Indeed, a few months after Hubble arrived at Mt. Wilson, Harlow Shapley, one of the more ambitious of the Mt. Wilson astronomers, journeyed east to Washington to defend the single galaxy theory against Heber Curtis.

Shapley, a talented astronomer, had achieved near celebrity status in 1918 by measuring the size of the Milky Way galaxy (which he believed to be the entire universe). Using the Cepheid variables and another type of variable star as standardized light sources, he estimated the Milky Way to be an enormous 300,000 light years across, ten times bigger than anyone had thought. We now know that Shapley was wrong by a factor of three: Our galaxy is approximately 100,000 light years in diameter. One reason for the miscalculation is that interstellar dust, not considered important at that time, may have been interfering with his viewing. In line with his belief that the Milky Way was the entire universe, Shapley claimed that the luminous nebulae were glowing clouds of gas located within the galaxy.

Hubble disagreed. In 1923 he discovered a Cepheid variable star in the Andromeda nebula, and using Shapley's own technique, found that the nebula was almost one million light years away, far beyond the outskirts of the Milky Way. Today we know that Andromeda is two million light years away. But Hubble's calculation established that the Andromeda nebula was in fact a full-fledged galaxy containing tens of billions of stars, much too distant to be part of or even associated with the Milky Way. Hubble did not publish his results for at least a year, checking and rechecking to make sure his conclusions were correct. He wrote to Shapley, now at Harvard, with the results of his work. Shapley argued against Hubble's conclusions but had to admit defeat in the end. Now, thanks to Hubble's careful work at the largest telescope in the world, it turned out that the vast majority of Messier's catalogued "nebulae" were actually faraway galaxies.

THE EXPANDING UNIVERSE

It was the Copernican revolution again. Not only was the Earth not the center of the universe, but the galaxy it was located in was merely one of millions in a universe more enormous than anyone had imagined before. Yet Hubble was not finished. By 1928 he had moved on to a new problem to solve that brought him into the heart of the static versus dynamic universe debate.

Astronomers had already noticed that the light emanating from many of the nebulae—now called "galaxies"—was redder than it "should have been." The most likely source of the redness was the Doppler shift toward the red end of the electromagnetic spectrum, meaning that the galaxies were moving away from observers on Earth. As he analyzed the photographic plates of forty-six galaxies taken by his assistant, Hubble knew the distances to half of them because of the Cepheid variables or novae within them.*

Based on their redshifts on the photographic plates, except for a couple of nearby galaxies including Andromeda, all of the galaxies appeared to be hurtling away rapidly from the Earth. Hubble soon noticed something remarkable: The farther away the galaxies were, the faster they were flying outward. In fact, there was such a linear relationship between distance and speed that if Hubble knew the speed of a galaxy, or the amount of redshift in its spectrum, he could easily compute its distance from the Earth. When dividing a galaxy's speed by its distance, the same number kept coming up: 150 kilometers per million parsecs per second, where a parsec equals 3.26 light years. This is a constant of proportionality linking the speed of recession of galaxies to their distance from us, and it soon came to be known as "Hubble's constant."†

The farthest galaxies from Earth are speeding away the fastest, and their speed is directly proportional to their distance. This surprising yet simple statement implies a couple of very interesting things. First, the universe is expanding. This was a momentous discovery. The data were showing that the universe was not static and eternal as Einstein and so many others had believed. If you run the clock backward, pulling the galaxies toward each

---

* A corroborating yardstick to the Cepheid variables was a type of supernovae. These exploding stars were assumed to have the same inherent brightness, so that calculating distances from their apparent brightness was quite easy.

† Its value has been revised downward several times since Hubble's day, until today, the best measured value by the Hubble Space Telescope is seventy-one kilometers per million parsecs per second.

other at rapid speeds, in a cosmic falling-in toward a center, that implies that the universe could have begun as a point, or a zero-volume of matter with infinite density at time t = 0. Lemaître's idea of the primeval atom must be correct!

Secondly, the proportionality constant allowed Hubble to work time into the equation. His calculations determined that the age of the universe was about 1.8 billion years. That is, the explosion that started the expansion of the universe took place 1.8 billion years ago. Hubble's calculation was far off the mark: According to modern observations and calculations, the universe is almost fourteen billion years old. But Hubble's calculation was a first approximation; it took several decades and new data to work out the universe's correct age.

Word of Hubble's discovery soon reached Einstein. Recall that against his better judgment, Einstein had inserted his fudge factor, the cosmological constant, to ensure that the universe would be static and eternal. Now, suddenly, the cosmological constant was unnecessary: The universe was in fact dynamic and expanding, just as Friedmann's solutions of Einstein's equations had allowed. This is the point at which Einstein declared that the cosmological constant was his "biggest blunder." When the Einsteins traveled to California in 1931, Einstein announced to journalists at Mt. Wilson that he had given up on his idea of a static eternal universe and felt that the model of an expanding universe was the correct description of the cosmos. Hubble's data had won him over.

Like Hubble's confirmation of the immense size of the universe, the concept of the expanding universe, with its implication of a violent birth, was a revolution in astronomy and cosmology. Like all paradigm shifts, it was not accepted quickly and easily by the scientific establishment.

For one thing, even though Einstein abandoned his cosmological constant, others such as Eddington and Lemaître still considered models that included it. Indeed today, in a revival of the idea, many cosmologists still use the cosmological constant, and view it as equivalent to the vacuum energy or dark energy. A serious criticism of the new expanding universe model was that the age of the universe as determined by Hubble's expansion law was only 1.8 billion years. This was more than three times younger than the age of the universe as determined by studying the evolution of stars, and about half the age of the oldest rocks discovered on the Earth. Attacking this perplexing problem, Eddington published a paper in which he suggested that the universe went through a "loitering" stage at its beginning. The effect of the cosmological constant in Eddington's equations for this model slowed down the expansion of the universe initially, thereby permitting an age of the universe compatible with the evidence obtained from stellar studies and geology.

Partly because of the age-of-the-universe problem, some cosmologists were seriously unhappy with the new expanding universe paradigm. According to Arthur Milne, an astrophysicist at Oxford University, the universe did not begin with the explosion of a primeval atom as proposed by Lemaître or a singular point as calculated by Friedmann, but rather, the rapid recession of the galaxies as observed by Hubble was due to *random motions* of the galaxies. The astronomer Fritz Zwicky thought the movement of the galaxies was an illusion. He proposed his "tired theory of light," in which the gravitational force of a galaxy in effect pulls back the light escaping from it, causing a redshift. His theory was soon shot down, though, because it could not explain the very significant redshifting that Hubble had discovered in the atomic spectra. Meanwhile Eddington, now won over to the new model as much as Einstein, continued promoting the idea of the expanding universe. In 1933 he published a popular book called *The Expanding Universe*, and said that in the future, larger telescopes would build on Hubble's observations and confirm the fact that the universe was indeed expanding.

FORGING THE ELEMENTS AND PREDICTING THE CMB

In today's universe, there is a highly skewed distribution of the abundances of the chemical elements. There are about ten hydrogen atoms for each helium atom. Other elements are represented in tiny amounts. For example, there are only six oxygen atoms and merely one carbon atom (the basis of life) to 10,000 hydrogen atoms. Other elements such as iron, uranium, mercury, and calcium are even less common than carbon. Would cosmologists be able to explain the evolution of the elements using the model of the expanding universe and the big bang?

It occurred to one creative physicist, George Gamov, a 1934 Soviet emigré to the United States, that perhaps the fusion process needed to create hydrogen from free electrons and protons, and then helium from hydrogen, would have started in the extreme heat at the birth of the universe. Gamov strongly disagreed with Lemaître's idea that the radioactive decay of the primeval atom was responsible for forming all the elements. He correctly understood that this would result in our seeing mainly the elements in the middle of the periodic table in the universe today, heavier elements such as iron, which is in blatant disagreement with reality.

Instead, and in contrast to Lemaître's primeval atom, Gamov proposed that the beginning of the universe was dominated by hydrogen and radiation.

He believed that it would be easier to obtain the fusion of hydrogen to helium if the early universe consisted mainly of the lightest element. Studying the work of physicists Friedrich (Fritz) Houtermans and Hans Bethe convinced him that the temperatures are not high enough in stars to fuse hydrogen atoms into helium. Hydrogen-to-helium fusion must have taken place at the origin of the universe when the intense heat could easily provide an efficient mechanism for producing the ratio of these elements that we observe in the universe today. Not only the ratio of helium to hydrogen, but also the actual amounts of the atomic constituents of these elements must remain the same throughout the history of the universe, after it is born and as it expands. This is now referred to as the "conservation of baryon number." That is, no baryons have been created or destroyed since the universe began.* Gamov worked back in time to the beginning of the universe, estimating the temperature of the original cosmic soup, and proved that at these extremely high temperature conditions, the abundances of hydrogen and helium observed in the present universe were produced naturally.

In 1944, Gamov took on a student named Ralph Alpher, a brilliant mathematician. After three years of work, Gamov and Alpher were able to show convincingly that helium was formed from hydrogen during the first few minutes after the universe was born. This was a very important result, because it boosted the whole idea of an expanding universe with an explosive origin. Gamov and Alpher were now able to account for more than 98 percent of all the atoms in the universe: specifically, the hydrogen and helium.

But there remained a vexing problem, and it was a blemish on the face of the expanding universe model: They were unable to explain the origin of the heavier elements. If through fusion we keep adding protons and neutrons to a nucleus, we reach a nucleus of five "nucleons," the term for both protons and neutrons in a nucleus. A five-nucleon nucleus is unstable due to the physics of the strong interactions holding it together. The unstable gap between four and six nucleons is like ascending a flight of stairs and finding that a step is missing. Gamov and Alpher had somehow to step over the unstable five-nucleon nucleus to reach the heavier elements. How? This question would not be answered for almost a decade.

Soon after World War II ended, Robert Herman joined Alpher in his research, and Gamov became mentor to the two talented young scientists.

---

* Baryons are subatomic particles made up of three quarks, and include the proton and neutron.

Together they improved on the calculations of the hydrogen-helium fusion process in the early universe, but they still ran up against the five-nucleon stumbling block. However, in the course of their research, Alpher and Herman made one of the most important predictions ever made in cosmology. In their vision of the birth of the universe, about 400,000 years after the explosion, the universe was "cool" enough (still a sizzling 3,000 degrees Kelvin) to allow the free protons and electrons to combine into stable hydrogen atoms, and to allow light to be released.* This harks back to Gamov's original idea that the very early universe consisted predominantly of hydrogen and radiation. This brief period Alpher and Herman christened with a misnomer. They called it the "era of recombination," when in fact this was the very first time that protons and electrons had combined to form hydrogen atoms. A more poetic name for this crucial time after the explosive birth of the universe is the "surface of last scattering." That is, until the protons and electrons combined, radiation was trapped, scattering crazily off the protons and electrons in the hot plasma and creating an opaque cosmic fog. Then, when things cooled enough, the protons and electrons combined to create hydrogen, which stopped the photons from scattering. Now the liberated light flowed freely, and the universe became transparent.

Here is the dramatic prediction that Alpher and Herman made in their 1948 paper: At that point when light became free-streaming throughout the universe, it would create a fossil of radiation in the heavens that could actually be detected today. The infrared light released about 400,000 years after the big bang would be redshifted significantly because space has stretched by a factor of over 1,000 since then. Therefore if we went looking for this radiation today, we would find it in the form of microwaves. This radiation—the cosmic microwave background radiation (CMB)—should be *everywhere.*

This was an immensely important idea, which would eventually lead to a modern confirmation of the expanding universe model with a sudden beginning. But it was far ahead of its time. Alpher and Herman were ignored by the cosmology establishment and no one thought about trying to detect the CMB radiation. The two young scientists both left academia in the mid-1950s to work in industry. It would be many years before the idea of the

---

* The Kelvin temperature scale, designed by Lord Kelvin (William Thomson) in the mid-1800s, provides a means of measuring very cold temperatures. Its starting point is absolute zero (0K), the coldest possible temperature in the universe, which corresponds to −273.15 degrees Celsius; the freezing point of water is 273.15K (0°C), while the boiling point of water is 373.15K (100°C).

CMB came up again, rediscovered by the Princeton University astronomer Jim Peebles.

THE STEADY STATESMEN

A strong challenge to the new, expanding universe model came from three British cosmologists in the late 1940s. Thomas Gold, Hermann Bondi, and Fred Hoyle decided to fight back against the expanding universe enthusiasts, who by then represented the majority of cosmologists, and they reintroduced the idea of an eternal universe, though with a somewhat dynamic twist.

Bondi, Gold, and Hoyle argued for the universe being in a steady state—and for that they were jokingly referred to in the cosmology community as the "steady statesmen." They proposed a quite attractive theory in which the universe could be eternal and essentially unchanging even while the galaxies were observed to be moving away from one another. They suggested that there was a continuous creation of matter in the voids between galaxies over time, and so, as galaxies moved away from one another, new ones continued to develop in the spaces that the older ones left. Thus the continuous birth of galaxies produced a "steady state," and there need not be a beginning to the universe.

This theory soon became a serious alternative to the expanding universe theory, for it agreed with all the key astronomical observations of the day. In particular it agreed with Hubble's redshift observations. Bondi, Gold, and Hoyle argued that the continuous creation of matter was no more inexplicable than the sudden appearance of the entire universe from a singular point. And so what if the steady statesmen were unable to explain the *cause* of the continuous creation of matter? The opposition was similarly unable to explain the *cause* of the explosion at the birth of the universe.

There were a couple of quite interesting aspects to the steady-state theory, including a testable prediction. For one thing, Gold extended Einstein's cosmological principle and postulated the "perfect cosmological principle": Not only is the universe isotropic and homogeneous, but also every period in time in the growth of the universe is the same as every other period.* He argued that the steady-state universe was a natural demonstration of this principle, and if the principle were really true, it would by definition rule out a moment of creation, whether by an explosion or any other means.

---

* Isotropic means that the universe looks the same to us when we look in every direction, while homogeneous means that it looks the same to every observer, anywhere in the universe.

Secondly, in order to describe mathematically the continuous creation of matter in between galaxies, Hoyle postulated a "C-field" ("creation" field). This field permeated the universe much like the old ether, and it spontaneously generated matter.[1]* Hoyle's equations required that only one atom be created every century in a volume equivalent to several city blocks—a very slow rate of creation, compared to the idea of a primal explosion!

Hoyle even proposed a way of testing the steady-state model. He said that new, small galaxies should be distributed equally throughout the universe, from our local cosmic neighborhood to the farthest reaches of space, since they would be popping up in the ubiquitous C-field. In contrast, the exploding universe model—in which matter was created at one moment at time t equal to zero (t = 0)—would not have uniformly distributed galaxies. The oldest infant galaxies would be the farthest away, and would presumably not even be visible beyond a certain age, because in the very early universe there were no fully formed galaxies. In the 1940s and 1950s, this was a purely academic issue, because no telescopes were powerful enough to see far enough back in time to vindicate one or the other of these two models.

Emotions were strong on both sides of the debate and sometimes the arguments became quite vicious, stooping to personal attacks. When Pope Pius XII endorsed the exploding universe model as proof of the existence of God, Hoyle, a well-known atheist, was incensed. He blasted the Vatican for using science and cosmology to prop up the evidence for God, and also for still not having "pardoned" Galileo for believing in the heliocentric universe.† Even the godless Soviet Union got in on the debate. Unwilling to promote the idea of a creator, they sided with the steady-state model.

It would be more than a decade before astronomers found the most convincing evidence to support the exploding universe model—or more properly, the dynamic evolving model—and defeat the steady statesmen. But in the meantime, Hoyle, who had a way with words, unintentionally gave the exploding universe model the name that stuck. In 1950, during a famous BBC radio program, *The Nature of Things*, while criticizing the new cosmological model, Hoyle referred to it dismissively as "the big bang."

---

* In hindsight, Hoyle's steady-state model has much in common with modern inflationary models. However, there is no need to introduce a creation field, for the inflationary period of the very early universe lasts only a very short time.

† It was not until 1992 that Pope John Paul II officially pardoned Galileo and admitted that Galileo and Copernicus had been correct in maintaining that the Earth moved around the sun.

RADIO ASTRONOMY SETTLES THE ISSUE

Since the 1930s, technological advances have expanded our research horizons so that telescopes are not limited to capturing visible light, but are now able to make use of the entire electromagnetic spectrum. Along with optical telescopes, we now have radio telescopes, infrared, and X-ray telescopes. During World War II, the invention of radar, developed from radio technology, helped England fight back against the Nazi bombing. After the war, some British wartime radar researchers created radio astronomy, using radar parts left over from the war.

Among them was Martin Ryle of Cambridge University. He invented the idea of using interferometry—in which several radio telescopes work together to resolve a signal more accurately—to perform astronomical observations. Ryle used his new tool to test the steady-state model versus the big bang model, by seeing whether young galaxies were only located far away or whether they also existed in our cosmic neighborhood. Fortunately, one characteristic of young galaxies is that they emit large amounts of radio waves.

Ryle's results, after surveying several thousands of radio galaxies, showed that these young galaxies were consistently found very far away from the Milky Way, thereby supporting the big bang model. Hoyle fought back, claiming that Ryle's observations were filled with errors and his statistical analysis was incorrect. But Ryle steadfastly defended his results, and stressed that his observations had been independently confirmed by an Australian group's radio astronomy data in the southern sky.

Another blow to the steady-state model came in 1963 when the American astronomer Maarten Schmidt discovered the first "quasi-stellar radio source" or "quasar" at the far side of the universe. Soon radio astronomers found more quasars, and their locations—all seemingly on the outer edges of the universe, suggesting that they were cosmic fossils from a much earlier epoch—favored the big bang theory and a dynamical, evolving universe. Soon cosmologists began to abandon the steady-state ship, and the arguments of the steady statesmen began to unravel.

WRAPPING UP THE NEW STANDARD MODEL OF COSMOLOGY

The final blow to the steady-state model came in 1964, when Arno Penzias and Robert Wilson, working at the Bell Labs Holmdel site, discovered the cosmic microwave background (CMB) radiation. This astonishing discovery, also made through radio astronomy, confirmed the prediction of Alpher and

Herman sixteen years before. The fossilized original light—the afterglow of light streaming forth at the surface of last scattering about 400,000 years after the big bang—had actually been found!* This clinched the whole idea of an original explosion and a finite beginning to the universe.

Fred Hoyle, who continued defending and tinkering with the steady-state model with new collaborators until the end of his life, actually contributed one of the last missing pieces of the big bang model: an understanding of how the heavier elements in the universe were formed. George Gamov had realized that hydrogen and helium originated in the primordial soup in the early universe. But what about the carbon that builds our bodies? Or the iron in the molten core of our planet? Hoyle suggested that all the elements heavier than helium were formed in the nuclear burning of stars long after the beginning of the universe, and he proposed the existence of an excited state, or "resonance," of carbon-12 that would decay to form the needed stable carbon nucleus to get this process going.[2] He stressed that since life exists and is based on carbon molecules, therefore the excited state of carbon-12 must also exist to have created the life-forming carbon.†

No such carbon resonance was known experimentally at the time, in 1953. Hoyle predicted that it would be found at 7.6 million electron volts in the nucleus of carbon-12. During a trip to Caltech, he tried to persuade the nuclear physicist William Fowler to search for this resonance state in his laboratory. Grudgingly, Fowler and his team performed the experiment and were very surprised to find the resonance just a few percent above Hoyle's prediction.[3]‡ Thus the heavier elements were not produced during the big bang, but were forged in the furnaces of stars as they formed, burned nuclear fuel, and died or exploded as supernovae, over and over again during many millions of years.

Although many people felt, as Einstein and Hoyle had, that a static, unchanging, eternal universe was beautiful and satisfying, the violent big bang model did solve one long-standing problem in a very satisfying way. As

---

* Now that we understand what it is, we can see the CMB anytime in a television set as the snowy picture issuing a constant hissing sound.

† This prediction of a carbon-12 excited state has been used recently by physicists advocating the "anthropic principle," that is, that the universe demands certain finely tuned constraints and physical events that lead to our own existence. See Chapters 6 and 8.

‡ For this work, Fowler was awarded the Nobel Prize in 1983, while Hoyle, who made the prediction, was ignored. The famous state of the carbon atom, however, was soon called the "Hoyle resonance."

early as 1610, Johannes Kepler wondered why the sky is dark at night. This question was most clearly stated by the German astronomer and physician Heinrich Wilhelm Olbers in 1823. He asked why, if the universe is infinite in size, is it not flooded with light? The question is now known as "Olbers' paradox." That is, if we existed in an infinite universe, there would be an infinite number of stars and galaxies emitting an infinite amount of visible light, and therefore our night sky would be very bright indeed. But the big bang model, now apparently proven to be true, represents a finite universe with a definite beginning and perhaps someday an end. Matter is not being continually created, as the steady statesmen had proposed. Assuming the universe had a beginning almost fourteen billion years ago, then the light from galaxies would only have had a finite amount of time to reach us here on Earth because of the finite speed of light. A finite universe can only emit a finite amount of light.

Today the common name among physicists for the homogeneous, isotropic, and expanding big bang model is the FLRW model, which stands for its chief inventors, Friedmann, Lemaître, Robertson, and Walker. After initial work by Friedmann and Lemaître, the American physicist Howard Percy Robertson and his English mathematician colleague Arthur Geoffrey Walker found exact solutions to Einstein's field equations that came to be known as the Robertson-Walker metric. However, the Robertson-Walker spacetime metric is independent of any dynamical gravitational theory such as Einstein's.

Evidence for the big bang model has continued to accumulate over the years, particularly from data gathered by the more recent Cosmic Background Explorer (COBE) and Wilkinson Microwave Anisotropy Probe (WMAP) satellite missions. These missions have provided even more details about the very early epoch of the universe captured in the cosmic microwave background radiation. However, as we shall learn, the possibility of an eternal or cyclic universe has come back into vogue among some contemporary cosmologists.

Chapter 4

DARK MATTER

A decade or so after Einstein's Nobel Prize and widespread acceptance of his radically new concept of gravity, some troublesome new data popped up on the motions of galaxies in galaxy clusters. These data did not fit Einstein and Newtonian gravity. Instead of questioning the theories, astronomers and physicists hypothesized the existence of exotic "dark matter" to explain the stronger-than-expected gravity they were seeing.

Today, like the elusive planet Vulcan in the nineteenth century, dark matter is accepted by the majority of astronomers and physicists as actually existing. Dark matter, although it has never been seen, is part of the generally accepted standard model of physics and cosmology, which also includes the big bang beginning to the universe.

We must take a look at the history of the discovery of galaxies and clusters of galaxies in order to understand what dark matter is all about and why it has become a major ingredient of the prevailing cosmological model.

DISCOVERING GALAXIES AND GALAXY CLUSTERS

On the night of April 15, 1779, the French astronomer Charles Messier was watching Comet 1779 as it appeared to pass slowly between the constellations Virgo and Coma Berenices on its journey through the solar system. On this night Messier found something in the sky that he was not looking for, a frequent occurrence in the history of astronomy. He was distracted from his observations of the comet by annoying, faint, diffuse objects that did not appear to move. He noted that thirteen of the 109 stationary patches that he identified were located where Virgo and Coma intersect. These objects comprise what is now known as the Messier catalogue of galaxies and supernovae remnants. Messier also found and named what we now know to be the Virgo cluster of galaxies, the nearest of the great clusters to us.

The brother-and-sister astronomy team, William and Caroline Herschel, took up the search for Messier's "diffuse objects" from their back garden in England a few years later. With the aid of improved telescopes, they found more than 2,000 of them, including 300 in Messier's Virgo cluster alone. The galaxies that Messier and the Herschels were observing looked diffuse or blurry through their telescopes because each was made up of millions of stars. Just as galaxies are assemblages of stars, galaxy clusters are collections of galaxies. In the hierarchical structure of the universe, galaxy clusters are only one level below the category of the universe itself. Thus in studying clusters, we are like observers viewing the universe from outside. The stars and galaxies in a cluster are of every age and type, providing a cross-section of the overall cosmic substance. Because a cluster is held together by gravity on a very large scale, its evolution, history and structure provide clues to the evolution of the entire universe.

In 1933, the American astronomer Harlow Shapley catalogued twenty-five galaxy clusters. By 1958, George Ogden Abell at the Californian Institute of Technology published a catalog of 2,712 clusters based on the Palomar Telescope Sky Survey. The revised Abell catalog now contains 4,073 galaxy clusters, with an additional 1,361 clusters from a survey of the southern sky supplementing the original northern survey. Today, all photographic surveys of the sky have revealed about 10,000 galaxy clusters—and the number keeps increasing with the aid of modern survey techniques and the Hubble Space Telescope. This is a measure of how amazingly large the universe is.

INVOKING DARK MATTER

The modern story of dark matter begins in the 1930s. The astronomers Fritz Zwicky and Sinclair Smith were studying the Coma and Virgo clusters, measuring the speeds of the galaxies within them by Doppler shifting. They discovered to their surprise that the galaxies were moving at speeds much greater than was supposed to be possible according to gravitational theory, if the clusters were to remain stable objects in the sky.

The greater the mass in a cluster, the greater will be the force of gravity holding it together. A cluster's mass is estimated by combining the masses of all visible galaxies and gas within it. The "escape velocity" is a theoretical calculation of the speed at which a galaxy at the periphery of a cluster would be able to overcome the gravitational attraction of the whole cluster system and literally escape its gravitational pull and fly out into space, much as a rocket overcomes Earth's gravity at a certain (escape) velocity.[1] Thus the escape velocity can be interpreted as an upper limit for the speed of moving galaxies that allows for a cluster's stability. Something was seriously wrong with the calculated escape velocities of the galaxies in the Coma and Virgo clusters.

According to Zwicky and Smith's calculations, the Coma and Virgo clusters should have dissipated long before their estimated lifetimes, losing their outer galaxies one after another as their speed overcame the cluster's gravity. To the two astronomers, it appeared that the mass of a cluster continued to expand linearly outward from its center, and in order to ensure the system's stability, its total mass would have to be many times greater than the mass estimated from visible galaxies and gas.

In order to get around this disturbing finding and explain a cluster's stability, Zwicky proposed that a halo of "dark matter" could exist around the center of the cluster, which according to Newton's law of gravitation would increase the gravitational attraction of the individual galaxies and therefore increase their escape velocities, leading to a cluster with the observed long lifetime. With this suggestion, Zwicky initiated the idea that there is "missing mass" in the universe and that it can be detected by its gravitational influence, even though it is not detectable by observations through telescopes. Zwicky did not suggest that there could be a flaw in Newtonian gravity.

CLUSTERS

The galaxy clusters are large objects in the sky, measuring many millions of light years across. Some are so large, in fact, that they have been called "superclusters." Because they are so huge, it is difficult to obtain accurate measurements of the motion of individual galaxies and thereby estimate their velocities. The galaxies in clusters are not uniformly distributed, either. They form what is known in physics as self-gravitating systems, in contrast to bound systems like the solar system and the galaxies, in which planets and stars, respectively, orbit a central mass. To analyze the motion of galaxies in a cluster, astronomers employ an averaging method called the "virial theorem," which determines the average speed of the galaxies from their estimated average kinetic and potential energies. This method is consistent with Newtonian gravitational theory for systems in equilibrium and works for many clusters to give a reasonable estimate of the velocities of galaxies and also the amount of dark matter needed to stabilize the clusters.

In 1970, a new satellite called Uhuru, meaning "freedom" in Swahili, was launched by the United States from Kenya. It began observing X-rays, a form of radiation not seen in astronomy before. The astronomers Edwin Kellogg, Herbert Gursky, and their coinvestigators at a small company in Massachusetts called American Science and Engineering obtained measurements from Uhuru of the Virgo and Coma clusters and found that they contained huge amounts of gas as well as galaxies. This gas, consisting of hydrogen and helium, is too thin to be seen in visible light, but due to its intense heat—more than twenty-five million degrees Celsius—it emits large amounts of X-rays. This hot gas in the clusters cannot be the missing dark matter, because it is detected through the X-rays it emits and thus is "visible" within the electromagnetic radiation spectrum. But measuring its mass is necessary when estimating the mass of the whole cluster.

As for measuring the stellar matter within clusters, we have to arrive at a number for the luminous mass of galaxies in the cluster. Thus "weighing" a galaxy amounts to estimating how much mass within it is producing a luminosity approximately equal to the sun's. Estimating a galaxy's luminosity is tricky because it depends on knowing our distance to the galaxy. Also, in any galaxy there are lots of aging stars that are dim compared to the sun, so a galaxy's mass may well be greater than it appears to be. Astronomers try to take such factors into account when estimating masses of galaxies.

There is a second kind of gas in galaxies and clusters—a *diffuse gas* that cannot be detected by radiation emissions, yet it must also be estimated in order to obtain the total mass of a galaxy or a cluster. Such a gas can absorb light from distant objects such as very bright quasars, thereby permitting an estimate of the mass density of the gas. Observational estimates of this material obtained by measuring atomic absorption spectra, as opposed to emission spectra, show that the mass of this diffuse gas is much less than the estimated luminous mass of the galaxy. On the other hand, the total mass of the hot, X-ray-emitting gas in the clusters is about two or three times more than the luminous mass associated with the stars.

Beginning in 1971, satellite measurements of the X-rays emitted by the gas in the Virgo, Perseus, and Coma clusters have allowed astronomers to determine the total mass required gravitationally to hold the gas in a cluster. The thermal origin of the X-ray emissions indicated that there is so much gas within clusters that they are actually huge "fuzz balls" of hot gas in the sky. The hydrostatic equilibrium of the galaxies bound by their self-gravitating attraction determines the total mass of the cluster. The hydrostatic equilibrium of a gas is determined by a Newtonian formula that balances the repulsion of pressure in an astrophysical body with the attractive pull of the mass of the gas. These X-ray observations again show that Newtonian gravity, together with the visible mass and the mass of the hot gas, is not consistent with the data, because the escape velocity of the galaxies is less than their greatest measured speeds, which would result in an eventual evaporation of the cluster.

Most physicists and astronomers now agree that to keep the laws of Newtonian and Einstein gravitation, there must exist in clusters a missing dark matter that is about *ten times more* than the combined mass of the stellar matter and gas. This is essentially the same conclusion that Zwicky and Smith arrived at in the 1930s. It indicates the enormous discrepancy between the predictions of the prevailing theory and what astronomers are actually observing.

GALAXIES

Not only galaxy clusters, but galaxies themselves are yielding similarly puzzling information leading scientists to invoke dark matter. The speed of stars orbiting the central mass of a galaxy can be determined accurately from the Doppler shift of the spectral lines emitted by the atoms making up the luminous mass and gas of the star. By measuring the shift in the spectral lines of stars toward the red or blue sides of the spectrum, we can tell whether a star is moving away or toward us, and even determine its speed as it moves around in its galaxy. The light emitted by stars moving away from us on the rotating side of the galaxy is observed by telescopes to be shifted toward the red end of the spectrum, while those moving toward us show a shift toward the blue end of the spectrum. By measuring these small shifts, in comparison to the much larger red shift due to what is called the "Hubble recession" of the galaxy in the expanding universe, astronomers can accurately determine the rotational velocity of the galaxy.

In 1970, the Australian astronomer Ken Freeman published a paper in the *Astrophysical Journal* claiming that his observed rotational velocities of stars were in disagreement with Newtonian dynamics. He concluded that there must be undetected dark matter equal to at least the visible mass of the galaxy.

Beginning in the 1980s, the American astronomer Vera Cooper Rubin and her coinvestigators did a more extensive survey of galaxies. Like Freeman, they discovered that the rotational velocities of spiral galaxies were faster (also termed "flatter") than they should be, according to the estimates of the velocities using the Newton-Kepler law of gravitation and the estimated visible mass of the disk or bulge of visible stellar matter at the cores of the galaxies. Moreover, in addition to this visible stellar matter, the galaxies contain neutral hydrogen and helium gases, which can be observed using infrared detectors, that is, detectors that measure the light emitted by the gas in the infrared frequency of the radiation spectrum.

To understand the problem of the too-rapid speed of the stars in a galaxy, picture an observer at the center of the galaxy looking through a telescope at stars moving around her in roughly circular bound orbits. In Newtonian mechanics, we say that the centripetal force on the star in its circular motion is balanced by the gravitational pull on the star by the mass at the center. According to the amount of *visible* stellar mass and gas at the center and Newton's law of gravity, the predicted speed of the rotating star is much less than the observed speed determined by the Doppler shift measurements. Following in Zwicky's footsteps for clusters, Rubin proposed that there must be a large spherical halo of missing dark matter around the core of a galaxy in order to account for the unexpected speed of peripheral stars that is observed.

Since Rubin's work, the rotational velocities of hundreds of galaxies have now been measured and cataloged. There is no doubt that there are only two ways to interpret the data: Either there is some sort of as-yet-unobserved dark matter creating a stronger gravitational field than there "should" be, or the gravitational laws of Newton and Einstein must be modified.

COSMOLOGY AND DARK MATTER

At the time of nuclear synthesis, which happened just seconds after the beginning of the universe in the big bang model, careful calculations show that, assuming Einstein gravity is correct, the amount of hydrogen and helium produced—accounting for 99 percent of the visible matter in the universe, in the form of protons or quarks—fits the cosmological data without modifying Einstein's gravity theory. Therefore, dark matter does not play a significant role in this very early epoch in the universe.

However, the universe expands, and by the time we reach about 400,000 years after the big bang, the whole picture changes dramatically. None of the recent accurate data for the CMB can be accounted for if we assume that the matter content of the universe is entirely made of visible baryonic matter. Therefore, the standard model of cosmology invokes dark matter to make Einstein gravity fit the data. As we shall see in Chapter 13, an alternative explanation modifies Einstein gravity. This allows us to fit the data with exactly the same number of protons as there were in the very early universe during the nuclear synthesis epoch.

SEARCHING FOR DARK MATTER

We can only verify the existence of any form of matter if we can "see" it with our eyes, with telescopes, or with instruments that detect photon emissions from different frequencies of the radiation spectrum. Since we do not see dark matter, we can only conclude that if it exists, it does not interact with light, or photons, as ordinary electrically charged matter does. Therefore, dark matter must only interact very weakly in collisions with ordinary matter such as electrons, protons, and neutrons, because otherwise we would already have detected it in our laboratories and from the spectral lines emitted by atoms in stars.

Hunting for elusive dark matter is now a multibillion dollar international scientific industry. Experimental physicists in pursuit of this mysterious dark matter hope to detect it eventually as it interacts very weakly with ordinary matter. Even though the dark matter remains invisible to the eye and the telescope, sophisticated experiments may be able to track its footprints as dark matter particles collide with the nuclei of atoms. This hope to detect a possible weak interaction of the dark matter particles with ordinary matter has spawned a large number of elaborate experiments. The projects require city-sized accelerators and underground detectors. Several teams in Italy, France, the United Kingdom, Switzerland, Japan and the United States are racing to capture dark matter particles. The Large Hadron Collider (LHC) at the giant particle accelerator at CERN (the European Organization for Nuclear Research) in Geneva, Switzerland, too, will soon begin experiments to search for the dark matter particles.

Many of the laboratories use high-purity material such as liquid xenon and germanium crystals cooled to low temperatures deep underground to shield the detectors from the continuous bombardment of ordinary matter particles that strike the Earth's atmosphere. The supposed weakly interacting dark matter particles would most of the time pass right through the detectors without hitting ordinary matter and thus remain undetected. However, experimentalists hope that on rare occasions a dark matter particle would collide with an atom. A recoiling of the atom's nucleus would cause a spray of electrically charged particles that could be detected.

There are several main candidates for the missing dark matter particles, and different laboratories are set up to detect them. One of the front-runners is WIMPs, "weakly interacting massive particles." WIMPs must weakly interact with nuclear matter and they must also have a large mass, at least several hundred to a thousand times more massive than the proton. Otherwise they would have already been observed in low-energy nuclear accelerator experiments. The WIMPs are called "cold" dark matter because massive particles move slowly in space and therefore their kinetic energy is not large enough to create a "warm" or "hot" material.

Another dark matter candidate is the axion, a by-product of the vacuum energy. In contrast to WIMPs, axions would have such a small mass that they, too, would not yet have been detected in experiments. The hypothesized "smallness" or "largeness" of the mass of the missing dark matter particles is synonymous with their ghostlike ability to elude detection.

Still another candidate for dark matter that was popular for a while is the neutrino, the only candidate to have been observed and is known to exist. Neutrinos, which have a small mass (although far bigger than the hypothesized axions), are fast-moving in space and are therefore referred to as "hot" dark matter particles. They can be considered a kind of dark matter because they interact only weakly with ordinary matter and they do not radiate photons. However, due to their small mass—less than two electron volts (an electron in these mass units has a mass of 500,000 electron volts)—and small estimated density, most investigators believe that they cannot account for the massive amounts of dark matter needed in galaxies and clusters of galaxies.

The neutralino is another intriguing possibility, and may constitute a kind of WIMP. In the popular supersymmetry theory of particle physics, the number of particles is obtained by doubling the number of known particles that have been observed in high-energy colliders. The shadowy doubles represent particles of opposite quantum properties from the known particles. So for example, in supersymmetry there is a photon and its partner the photino. One of these supersymmetric particles called the neutralino is a massive partner of the electrically neutral neutrino. In the overall zoo of possible dark matter particles, the neutralino is considered a strong WIMP candidate. Scientists at CERN hope to find it with their new LHC.

Because of the proliferation of dark matter candidates, theorists have great difficulty telling experimentalists precisely where and how a dark matter particle can be detected and whether, indeed, a detection of "something" actually constitutes the discovery of dark matter. Because the dark matter particles have never been detected experimentally, and yet are believed to be much more numerous than particles of visible matter, theorists know that they are either very light particles or that they are very massive, but theory does not tell us how light or how massive they should be.

Theorists are attempting to use a particle-physics version of supersymmetry theory called "supergravity" to predict properties of the dark matter particles. Supergravity unifies gravity with the electromagnetic and radioactive weak forces (called the "unified electroweak force") and the strong force that binds quarks, the basic constituents of the proton and neutron bound together in the nuclei of atoms. Supergravity is one version of the long-sought unified theory of all the forces of nature. In the language of particle physics and quantum theory, the gravitational equivalent of the massless photon is called the "graviton" and in supergravity it has its supersymmetric partner, the "gravitino." Basing their calculations on the many unobserved superpartners in supergravity, theorists attempt to provide some estimates of the masses of superpartner WIMP candidates that could be detected at the CERN LHC collider. Not only do the experimentalists have to detect a dark matter particle candidate, but they also have to verify that it is stable. This can only be done by observing its behavior in astrophysical systems and in cosmology.

Thus a considerable amount of theoretical guesswork goes into calculating the possible likelihood of observing the WIMPs in the big machines, and hunting for the dark matter candidates becomes the proverbial searching for a needle in a haystack. But what if there are no dark matter needles in the cosmic haystack at all?

It may be that ultimately the search for dark matter will turn out to be the most expensive and largest null result experiment since the Michelson-Morley experiment, which failed to detect the ether.

Chapter 5

CONVENTIONAL BLACK HOLES

I n 1979, Walt Disney released the science fiction movie *The Black Hole*. In the film, the spaceship *Palomino* discovers a black hole with another spaceship at rest outside the event horizon. That ship is the USS *Cygnus*, named after the first discovered double star system that supposedly contains a black hole. The *Palomino*'s captain, Dan Holland, and his crew board the *Cygnus* and meet Dr. Hans Reinhardt, a well-known scientist who has been missing for twelve years, and his crew: human beings whom Reinhardt has lobotomized and turned into semirobots. While all are onboard together, an asteroid storm wrecks the *Cygnus*'s antigravity generator, which has been maintaining the ship at a resting position near the black hole event horizon.

As the spaceship passes through the black hole event horizon, those on board do not experience anything unusual, such as a sudden jolt or explosion. This is exactly what Einstein gravity predicts will happen when observers fall through a black hole event horizon: It will feel perfectly normal. On the other hand, a distant observer would see the *Cygnus* spacecraft slowing down as it approaches the black hole, but never actually reaching it. This is because, to the distant observer, the visible light emitted by the spacecraft becomes infinitely redshifted the closer it gets to the event horizon.

The risible part of the Disney film plot is that as the *Cygnus* spacecraft with its *Palomino* survivors passes through the event horizon and is drawn in toward the central singularity of the black hole, the enormous gravitational tidal forces would stretch them out like rubber bands. Instead, the Disney spaceship neatly avoids this gruesome end by crossing over into another dimension or another universe, for Dr. Reinhardt has preprogrammed the *Cygnus* to enter the black hole seeking another universe. We watch the evil Dr. Reinhardt, split into two pieces, floating down through a corridor of cathedral arches for eternity. In a true black hole, with an infinitely dense singularity at its heart, the superstrong gravitational forces at the center would ultimately destroy anything unlucky enough to find itself inside.[1]

HOW DOES A BLACK HOLE FORM, AND WHAT DOES IT DO?

The concept of black holes arises from a solution of the basic field equations of Einstein's gravity theory, as worked out by Karl Schwarzschild in 1916. Consequently, black holes, like the big bang and dark matter, are part of the standard model of gravity and cosmology. According to general relativity, when the mass of a star is greater than three solar masses, then conventional nuclear physics cannot prevent the star from collapsing under its own gravity to zero radius, that is, to a black hole.*

According to the Schwarzschild solution of Einstein's field equations, first an event horizon forms as the star collapses and then all the matter of the star inevitably collapses through the event horizon toward the center of the star and forms a singularity with infinite density. An event horizon is like a door into a dark room that only lets you pass through one way. Once you are inside the room, you cannot ever return to the event horizon door and leave. This event horizon occurs when the radius of the collapsing star equals twice Newton's gravitational constant times the mass of the star divided by the square of the speed of light. This critical radius is called the "Schwarzschild radius." For the sun, the Schwarzschild radius is three kilometers, and if the Earth were to collapse, this critical radius for the event horizon would be only about one centimeter. The event horizon prevents any form of matter, including light, from escaping, and it characterizes a black hole. It is one of the great mysteries of modern physics that Einstein's field equations predict this extraordinary object in nature.

Einstein's general relativity actually predicts three kinds of black holes, depending upon the state of the original star: the famous Schwarzschild solution black hole with a mass M just described, the 1916–1918 Reissner-

---

* When smaller stars die, they become other celestial objects. If a star is less than 1.5 times the mass of the sun, it will collapse into a white dwarf, which is the fate of our sun. If it is between 1.5 and 3 solar masses, it will become a neutron star. The well-known Indian astrophysicist, Subrahmanyan Chandrasekhar, calculated these mass limits in 1930 while on a steamship bound for England to visit Sir Arthur Eddington.

Nordström solution, where the black hole has electrical charge as well as mass, and the 1963 Kerr solution with not only mass and charge but also angular momentum, which means that the Kerr black hole, like the star it came from, is rotating.

Astrophysicists also believe that black holes come in three sizes: mini, medium, and enormous. Mini black holes are believed to be primordial, created in the early universe; medium-sized black holes can form as a result of the collapse of a star in a binary star system; and enormous black holes are believed to lurk at the centers of galaxies.

Many physicists believe that mini black holes—with a mass roughly one-tenth that of the Earth—were formed during the big bang. The Schwarzschild radius for a mini black hole is about one-tenth of a centimeter or less, about the size of a pinhead.

The medium-sized black holes originate in a binary or double star system, which consists of two stars orbiting one another in close proximity. One of the stars can be at the stage in its life cycle when its nuclear fuel has burned out. The star explodes, and then its core implodes, forming a neutron star. The neutron star begins to accrete matter from the larger companion star and eventually the mass of the neutron star exceeds the Chandrasekhar mass limit for neutron stars, 1.5 to 3 times the mass of the sun. At this stage, the repulsive pressure of the degenerate neutron gas can no longer counterbalance the pull of gravity, and according to the equations of Einstein's gravity theory, the compact star undergoes further gravitational collapse to a black hole.

The third kind of black hole, occurring in the centers of galaxies, is an exceedingly massive object, from one million to one billion times the mass of the sun. Black hole sleuths call these supermassive objects at the centers of galaxies "galactic nuclei," comparing them to the heavy nucleus at the center of an atom. It remains a mystery how these supermassive black holes ever evolved at the centers of galaxies.

BUT DO BLACK HOLES REALLY EXIST?

Of course, black holes are never directly seen, so their masses cannot be calculated on the basis of their luminosity, which is one of the common methods for measuring the mass of a star. Therefore astronomers who claim to "see" a medium-sized black hole infer its mass from the motions of its optically visible binary companion, and attempt to observe theoretically predicted signatures such as the emission of X-rays from the accreting gas surrounding the black hole. X-rays are associated with black holes because the accreting gas heats up to temperatures that allow the emission of radiation in the X-ray part of the spectrum. The huge gravitational force at the event horizon that is sucking in the gas is what causes the extremely high temperature of the accretion disk.

There are a few popular black hole candidates, such as the Cygnus X1 binary system. This has a known dark companion star whose mass is estimated to be seven to ten times the mass of the sun, therefore exceeding the Chandrasekhar mass limit. At this mass limit, the star collapses and forms a black hole according to the predictions of Einstein's gravity theory. Since the event horizon is not a material surface such as that of an ordinary star, and any material falling into the black hole does not heat up further as it would on a material surface, we would not expect to see a glowing material surface at the black hole. But that very lack of a glowing surface, combined with the emission of X-rays at the event horizon, is considered a positive signature for double-star black holes. However, the calculations supporting the existence of this phenomenon are complicated, involving difficult, model-dependent astrophysical effects.

Black holes have caught the imagination of the media and the public. Most astrophysicists believe that black holes, as described by Einstein's gravitational theory, do exist, even though the evidence for them is circumstantial at best. Yet whether or not binary black holes really exist constitutes an ongoing and intense debate. Indeed, some astrophysicists claim that it is impossible to *ever* directly observe the event horizon of a black hole. The number of binary black hole candidates that astronomers have found is surprisingly small. In view of the trillions of binary star systems in the universe—roughly *half* of all the stars in the cosmos, in fact, are binaries—the heavens should be abundantly populated with binary black holes. But this does not appear to be the case.

The evidence for black holes at the centers of galaxies is also circumstantial. As in the case of the binary black holes, observers using, for example, the Hubble Space Telescope attempt to weigh the object at the center of the galaxy; they try to observe certain unique signatures such as X-rays emitted by the opaque gas surrounding the event horizon of the black hole. In addition, astronomers can observe the speed of stars orbiting some distance from the event horizon. The ideal signature of a massive black hole at the center of the galaxy would be to see objects disappearing into the event horizon, much as you can see ships disappearing over the horizon on a sunny day when standing on a beach looking out to sea. But the fact that stars are disappearing into the event horizon can only be inferred by observing the behavior of the dust, gases and stars at a significant distance from the black hole event horizon. In fact, a distant observer can never see anything crossing the event horizon: The in-falling stars will appear "frozen" as they approach it. The interpretation of observations near the supposed event horizon of a black hole depends upon which of several models one uses to describe the motion of the stars in the gas at the black hole event horizon.

The purported black hole at the center of our galaxy is located in the southern constellation Sagittarius ("The Archer"). It is about 26,000 light years away from Earth and is estimated to have a mass of about 3.7 million solar masses. The Schwarzschild radius of the event horizon for a black hole equivalent to 3.7 million solar masses is about ten million kilometers, or thirty-three light seconds, across. This is not as enormous as one might think. Compare the black hole's radius to the distance between the Earth and the sun, which is 150 million kilometers, or about eight light minutes across.* However, because of the huge opaque curtain of gas, many light years across, that surrounds the event horizon of the putative black hole at the center of our galaxy, we are unable to see the event horizon directly with optical telescopes in the range of visible light. On the other hand, by using giant telescopes in the infrared part of the radiation spectrum, we can obtain images of the region.

These images show a dense conglomeration of starlike objects, each with a mass less than or about equal to the sun. They appear to be compact stellar systems, which could be neutron stars, brown dwarfs, or mini black holes that were created in the early universe. Among this conglomeration of objects at the center of the galaxy could be the supermassive black hole.

Astronomers can tell, from studying the high resolution images of the motions of these many stars, that a mass of about three million times that of the sun is located within a radius of only ten light days or 2.6×10^{11} kilometers. This location is identified with the compact X-ray or radio source located in the center of the constellation Sagittarius A. The accurate telescopic images show the orbit of a star called S2, which circles Sagittarius A* (SGR A*; the asterisk means "star") with an orbital diameter of approximately two light days or 5×10^{10} kilometers, which is 5,000 times the Schwarzschild radius away from the supposed black hole. Astronomers infer from the rapid orbital motion of this star in the gravitational field produced by the dense object in Sagittarius A that only a black hole could produce such an orbit.

---

* We measure astronomical distances by the time it takes light, traveling at 300,000 kilometers per second, to cover them: one light hour equals 1.1×10^9 kilometers, one light day equals 2.6×10^{10} kilometers, one light month equals 7.8×10^{11} kilometers, and one light year equals 9.5×10^{12} kilometers.

FIGURE 6: **The black hole at the center of the Milky Way galaxy is in the constellation Sagittarius.** Source: Chandra X-Ray Center, Harvard-Smithsonian Center for Astrophysics.

However, others believe that further accurate observations are needed to determine whether the large, superconcentration of stars at Sagittarius A constitutes *in itself* the dense massive object inferred from the observations, or whether a genuine black hole lurks unseen in the background.

While most astrophysicists believe that the observed great speed of stars surrounding the supposed event horizon at the center of the Milky Way is sufficient evidence of the strong gravitational pull of a giant black hole, a vocal minority questions whether a unique signature associated with black holes has ever been observed.[2]

THE WEIRD WORLD OF BLACK HOLE PHYSICS

Whether black holes exist or not, they provided a needed boost that revitalized gravity studies when I began my physics career. By the 1950s, interest in working on Einstein's gravity theory was flagging in the physics community. There was a significant decline in the number of published papers on general relativity. Proof for the theory rested on just the anomalous perihelion advance of Mercury, Eddington's and subsequent solar eclipse observations of the bending of light, and gravitational redshift observations of white dwarf stars such as Sirius, which were inconclusive. Most young physicists preferred to work in the flourishing fields of nuclear physics and particle physics, which were very active areas for experimental research, rather than poring over complicated gravitational equations that had little hope of being tested or producing new physics. Indeed, when I entered the PhD program at Trinity College in 1954, I was the only student at Cambridge working on Einstein's theory—a situation that one could find at the major universities across Europe, Britain, and North America.

Then the black hole arrived. Groundbreaking papers that perked up physicists' interest in general relativity were published in the late 1950s. Based on earlier work by Eddington, in 1958 the American physicist David Finkelstein introduced coordinates that extended the spacetime of the original Schwarzschild solution so that one obtained a different understanding of the nature of the black hole's event horizon. Then the Norwegian-American physicist Christian Fronsdal and the Australian-Hungarian mathematical physicist G. Szekeres, in 1959 and 1960 respectively, provided additional important perspectives on the nature of the Schwarzschild solution.

Most important, also in 1960, the Princeton mathematician Martin Kruskal published the most complete description of the Schwarzschild solution by introducing what is now called the "Kruskal coordinate transformations." Among other things, the paper cleverly proved that there were not two singularities in black holes, as were originally thought to exist in the Schwarzschild solution, but only one. In the Kruskal coordinates, there is no real singularity at the black hole event horizon, but the essential singularity at the center of the black hole remains. Kruskal's discovery revived interest in black holes and therefore in general relativity. If the event horizon surrounding the black hole had continued to be thought of as a singular surface with infinite density, then this would have been unacceptable to even the most enthusiastic black hole protagonists.

In the 1960s and 1970s, increasing numbers of physicists, some famous, became involved in black hole research. Roger Penrose proposed what is now called the "Penrose diagram," which was an important mathematical understanding of the Kruskal completion of spacetime. The whole spacetime of the Schwarzschild solution became reduced to its essential details. Infinite distances were crunched into the diagram by what is called a "conformal mapping of coordinates." This mathematical description of the Schwarzschild spacetime allowed for a much clearer understanding of the black hole.

In the mid-1970s, Stephen Hawking began pondering whether whatever is swallowed by a black hole—a volume of Shakespeare, a spaceship, a star—is forever lost, or whether it can leak out over time. This is the famous "information loss paradox." Simply stated, the paradox holds that in a black hole, information disappears, and yet the laws of physics prevent informa-tion from disappearing.* Basing his arguments on the Israeli physicist Jacob Bekenstein's work on the thermal properties of black holes, Hawking showed that black holes were not really completely "black," but emitted radiation. This "Hawking radiation" is not ordinary radiation, but results from a quan-tum mechanical phenomenon taking place at or near the black hole's event horizon, and eventually it will radiate away all the matter and energy inside the black hole. Will there be any remnant left of the information inside the black hole when it completely evaporates, any ashes from the volume of Shakespeare?

Even posing this question is at odds with the essential singularity at the center of the black hole. It is hard to understand how this infinitely dense singularity can evaporate into nothing. For matter inside the black hole to leak out into the universe requires that it travel faster than the speed of light. Moreover, the matter must effectively go backward in time. According to the Schwarzschild black hole solution of general relativity, the concepts of space and time are interchanged within the black hole. We can reverse direction in space but we cannot reverse direction in time. We can only move into the future, not into the past. According to the special theory of relativity, noth-ing can move faster than light, so it is impossible for information to escape the black hole.

However, the Hawking radiation can allow the matter inside the black hole to leak out as *completely scrambled information,* that is, zero informa-tion, over a very long period of time. Completely scrambled information is like a wartime code, an encryption of secret information, for which the key that decodes it has been forever lost. How can we simply lose all the informa-tion in the black hole when the black hole evaporates? The information loss paradox contradicts the basic laws of quantum mechanics. One way to avoid it is to attempt to modify quantum mechanics. However, this is not a solution that is favored by most physicists because quantum mechanics agrees with all experiments to date.

The physicists Gerard 't Hooft and Leonard Susskind disagreed with Hawking's assertion that information would be lost once it fell through the black hole event horizon. They felt that Hawking had identified a serious paradox with deep consequences for gravity and quantum mechanics. They could not accept the idea that information was not conserved in the universe;

---

* Physicists use the term "information" from quantum mechanics in reference to black holes, rather than the classical term "matter." As with the conservation of energy and matter, we also speak of the conservation of information.

the conservation of matter and energy was too fundamental a tenet of physics to be discarded if the black hole paradox was true. 't Hooft and Susskind attempted to resolve the information loss paradox by introducing the idea of nonlocality into black holes. In classical physics, a particle has a definite position in space at a given time. The same is true in quantum mechanics once a measurement is made of an electron's position in space. But in the case of the black hole, the observer inside the black hole sees the information at a definite location, while the observer outside the black hole sees that same information in a different place, radiated back out as Hawking radiation from a region just outside the event horizon. If black holes are indeed real objects in the universe, then the concept that information has a definite location in space must be wrong. When objects do not have a definite position in space, that phenomenon is called "nonlocality." 't Hooft and Susskind promoted the idea of the hologram as a prime nonlocal object.

The idea is that all the information in the black hole, such as the unfortunate spaceship, is stored on the two-dimensional walls of the black hole, just as one would create a hologram. The perceptions of the two different observers in black hole complementarity—one inside and one outside the black hole—are simply two different constructions of the identical hologram using two different algorithms. The scrambled information encoded on the black hole hologram can be decoded by a quantum mathematical algorithm by shining a laser light on the hologram! The first reconstruction of the hologram shows the information perceived by the observer outside the event horizon, while the second reconstruction is what the observer falling through the event horizon sees. Two reconstructions of the same hologram produce two complementary pictures of the black hole scene.

For three decades, Stephen Hawking was convinced that information was lost and destroyed during the long period of time during which a black hole evaporates. In 1997, he and Kip Thorne, a physics professor at the California Institute of Technology in Pasadena, made a bet with John Preskill, also a Caltech professor, that information was truly lost in a black hole. They agreed to bet upon an encyclopedia from which the information would or would not be lost—although how they were supposed to determine the correct answer, and therefore the winning side, was never clear. In 2004, Hawking showed up unexpectedly at a relativity conference in Dublin, turning the gathering into a media event. During his formal talk, Hawking recanted. The electronic voice issuing from his high-tech wheelchair intoned that he had been mistaken: Black holes do not destroy information. He was as good as his word, publicly presenting Preskill with an encyclopedia of baseball.

Confusion ensued, for Hawking did not enlighten anyone about what had changed his mind, whether it was the hologram idea or something else. A year later, he published a seven-page paper on the electronic archives explaining his conversion. Many in the physics community, including myself, still do not completely understand why Hawking changed his mind about the information loss paradox. Most physicists are still in a quandary over whether information is really lost in a black hole.

In 2004, Preskill and the young quantum information physicist Daniel Gottesman published a paper called "The black hole final state." Gottesman is a colleague of mine at the Perimeter Institute for Theoretical Physics in Waterloo, Ontario, where I have been a researcher since my retirement from the University of Toronto. The Gottesman-Preskill paper explained the information loss conundrum by using the concept of quantum entanglement. A charged particle falling through the black hole event horizon gets quantum-entangled with another charged particle just outside the event horizon. Through the spooky nonlocality of quantum mechanics, the two particles are able to communicate instantaneously about their respective properties, such as quantum mechanical spin, without violating special relativity.* Gottesman and Preskill's conclusion was that the whole issue of the loss of information in a black hole was even more serious than Hawking and Thorne, prior to the baseball encyclopedia bet, had claimed for three decades. They did not agree with Susskind and 't Hooft's ideas on the hologram and black hole complementarity, but came down on the side of Hawking's original position that all information is lost in a black hole as Hawking radiation evaporates it away to nothing—or evaporates it away to scrambled information, which amounts to nothing. The quantum Hawking radiation would make the black hole evaporate in a finite amount of time, but this would contradict Einstein's gravity theory, which says it takes an infinite amount of time to form a black hole; the black hole would disappear before it came into existence.

DO WE NEED BLACK HOLES?

Is the reader feeling confused about the status of the black hole information loss paradox and black holes in general? So am I! Every time I have a discussion with another physicist about the information loss problem, an argument ensues, and no consensus is reached. No one really understands the informa-

---

* Although "observers" at great distance cannot communicate with those inside a black hole at the classical level of physics, it may be possible for quantum entangled particles to communicate through this barrier.

tion loss paradox. Perhaps the best way to get rid of this and other black hole problems is simply to get rid of the black hole! It helps to remind ourselves that Einstein, Eddington, and even Schwarzschild were all unhappy with the black hole solution, for they considered it unphysical—that is, they felt it was a mathematical artifact that surely did not actually occur in nature. But the only way to get rid of black holes is to turn to another gravity theory, one in which the field equations do not produce the singularities that result in black holes. Chapter 14 discusses a possible universe without black holes.

However, most physicists accept black holes, with all their problems, as part of the true picture of the universe, even though all evidence for them is circumstantial. Some would even go so far as to say that black holes may have practical value for research! For example, serious physicists have suggested that we as outside observers could possibly detect the formation of mini black holes from strings or some other form of exotic matter in the Large Hadron Collider (LHC) at CERN. Experimentalists dream of some spectacular discovery such as proof of the existence of black holes to justify the more than eight billion dollars it has cost to build the LHC. Others even suggest that one could use black hole physics from a higher dimension to investigate the properties of heavy ions in the heavy ion experimental research performed at Brookhaven National Laboratory.

PART 3
UPDATING THE STANDARD MODEL

Chapter 6

INFLATION AND VARIABLE SPEED OF LIGHT (VSL)

The original standard big bang model of cosmology has been remarkably successful in clarifying the first few moments of the universe. The isotropy in the model tells us that the temperature of the cosmic microwave background after the big bang is uniform in all directions, and the homogeneity means that on large scales, the universe appears the same to any observer anywhere within it. The model provides a description of nuclear synthesis, when hydrogen fuses to make helium and deuterium about one minute after the big bang. It predicts the abundances of the lighter elements in the universe. When Hubble found that all the distant galaxies were receding from one another at a speed proportional to the distance separating them, the big bang model was able to accommodate this startling observation.

However, there are some serious flaws in the big bang model, and this has motivated the search for other possible mechanisms operating in the fractions-of-seconds-old universe. These flaws are collectively known as the "initial value problem," which arises when we try to explain how the universe began, and what the values of different cosmological parameters were at that time. Stephen Hawking and Barry Collins first called the big bang scenario into question in a paper in the *Astrophysical Journal* in 1973. They showed that the standard big bang model had a probability of close to zero to explain how we got from time t = 0 to the present epoch, with the critical density of matter necessary to form our spatially flat universe! The big bang model was about as likely to explain this phenomenon as a pencil would be able to balance on its sharp tip. This major problem with the big bang model is referred to as the "flatness problem." In other words, if the universe was curved at time t = 0, then we have to explain how after fourteen billion years, it becomes approximately flat, as is observed today.* And so the question is: How do we get from the assumed curvature of space in the very early universe to the present *lack* of spatial curvature in our universe that has been revealed by cosmological data? The big bang model would have to be ultra-fine-tuned to achieve the matter density that is equal to the critical density necessary to produce the spatially flat universe that we inhabit today. This fine-tuning, a mathematical cancellation of very large numbers to up to sixty decimal places, is considered unacceptable by most physicists because it would bring about an unnatural initial state at the beginning of the universe.

A second serious issue is the "horizon and smoothness problem." The consensus among physicists is that in order to achieve uniformity in the early universe, whether of temperature or of density of matter, there must be "communication" among the parts of the whole; observers must be able to "see" or interact with each other and adjust their parameters in tandem. Milk poured into coffee will eventually lighten the liquid uniformly whether one stirs it or not. A blast of cold air through an open front door will within a few minutes lower the temperature uniformly everywhere in the room. An evolving infant universe will become homogeneous in a similar way, achieving a fairly uniform temperature distribution—as long as its protogalaxies and stars can communicate causally with each other. But according to the big bang model, this cannot happen in the early universe, for each part of it can only communicate with other parts at a rate limited by the speed of light. Within the first few seconds of the universe, there has simply not been enough time for light to have traveled from one part of the new, expanding universe to distant areas, and each protogalaxy is restricted in its line of sight by the blinders of its own horizon. Yet Arno Penzias and Robert Wilson's cosmic microwave background radiation (CMB) data in the early 1960s showed that there was, indeed, uniformity and homogeneity in the early universe because the microwave temperature is remarkably uniform no matter where one points one's radio telescope toward the sky. We just have no explanation in the big bang model of how the temperature got that way.

INFLATION TO THE RESCUE

In 1981, the MIT physicist Alan Guth proposed a modified version of the big bang model. Borrowing from the field of economics, he called it the inflation-

---

* Robert Dicke of Princeton University coined the phrase "flatness problem" in 1978. "Spatially flat" means that the geometry of three-dimensional space is Euclidean.

ary universe model, or simply "inflation." This new scenario neatly took care of the flatness, horizon, and smoothness problems of the standard big bang model. The Russian astrophysicist Alexei Starobinsky had published a similar but less well-developed idea of an exponentially inflating universe in 1979 and 1980; so had the Belgian physicists François Englert, Robert Brout and Eduard Gunzig in 1978, as well as Katsuhiko Sato in Japan. Starobinsky developed his model to remove the singularity at the big bang without claiming to solve the horizon and flatness problems. It was only after 1984 that it was recognized that his model could be adapted to inflation. Englert, Brout and Gunzig had part of the inflationary idea.

In a nutshell, inflation made the spacetime of the early universe expand enormously to about 10^{50} times its initial size in just a tiny fraction of a second. This exponential inflation, taking place almost immediately after the big bang, stretched spacetime beyond the visible horizon. This, according to Guth, could explain why the universe we observe and live in is so large, spatially flat, smooth, and uniform.

Consider the universe at the tender age of less than a million years old. If there were two observers in it, separated by a million light years, they could never communicate with each other or even know of each other's existence, due to the currently measured finite speed of light. In that case, how could the temperature of the universe become so uniform that the cosmic microwave background is observed to be the same in every direction, like a well-mixed cup of coffee? Back in 1981, Guth realized that an exponential expansion of the universe in its first seconds after t = 0 would make the universe inflate instantly, and like latex expanding, it would smooth out any wrinkles. This inflation causes horizons everywhere in the new universe to expand enormously, so that one hypothetical observer could now communicate the uniform temperature in her part of the universe to another observer far away, thus solving the smoothness problem. In other words, radiation from one part of the universe could reach other parts far away, leading to temperature equilibrium. Guth also showed that this new scenario could lead to a solution to the flatness problem because the sudden inflation of spacetime would erase its curvature and remove the need for the unacceptably extreme fine-tuning that was necessary in the big bang model.

Einstein's mysterious cosmological constant, or the vacuum energy, also comes into play in the very early universe. The vacuum energy is a repulsive force. In the inflationary scenario, it is what overcomes the force of gravity and drives the sudden inflationary expansion of spacetime. Cosmologists now view the vacuum energy as a constant property of empty space, so that as spacetime expands, the density of the vacuum energy remains the same throughout the universe. (The density of ordinary matter dilutes as the surrounding space expands and the universe ages.)

In the inflationary scenario, the cosmological constant is actually many orders of magnitude larger in the early universe, so large that the universe would double in size in just a tiny fraction of a second. In far less time than the wink of an eye, the universe would grow from the size of a proton to a size enormously larger than the universe we see today. Objects in the universe would soon be receding from one another at speeds exceeding the speed of light beyond an observer's visible horizon. According to this inflationary scenario, the superluminal (faster than light) expansion only happens beyond the visible horizon, so that an observer within his or her own horizon would not even be aware of it.

By definition, the cosmological *constant*, or the vacuum energy, is constant. How then, going along with inflation, can we visualize that it was enormously larger fourteen billion years ago than its presently tiny observed value? There must have been an incredibly large pressure of antigravity in the very young universe.

Some inflationists ask us to engage in a little mathematical mind-bending analogy. For this, we need to make use of the hypothetical inflaton particle. Just as the photon and graviton carry the forces of electromagnetism and gravity, respectively, some imaginative physicists propose an actual particle

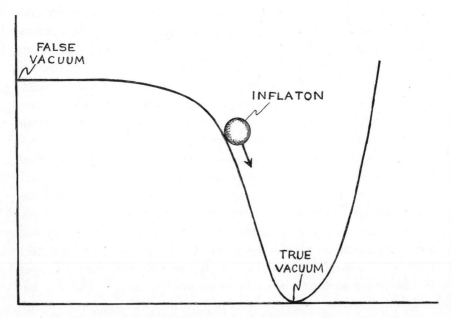

FIGURE 7: **The inflaton particle rolls down from the false vacuum to the true vacuum.**

that drives inflation. Picture the very early universe, or the inflaton particle itself, as a ball rolling along in the mountains, looking for a valley it can roll into, seeking the bottom of the valley where it reaches its minimum of potential energy. At its maximum potential energy on the mountain peak, the ball experiences a large, repulsive vacuum energy. This makes the ball-universe undergo a huge inflationary expansion, and then when the ball rolls into its minimum potential energy at the bottom of the valley, the inflation ends.

An unknown mechanism makes the ball roll very slowly along a flat plateau, in its search for a valley, with the vacuum energy varying very little. Physicists call this plateau the "false vacuum," and it represents the period of inflation. If this slow roll takes long enough, the universe would increase vastly in size before the ball begins a steep descent into the valley where inflation ceases. When the ball gets close to the bottom of the valley, the universe flattens out and appears uniform, just as the CMB data show. If the cosmological constant or vacuum energy is small enough, with a near-zero altitude above the bottom of the valley, then our universe can come into being, and galaxies, stars, planets, and life can evolve, now that the inflationary period is over. How and why the cosmological constant today has the very tiny value needed to allow for the evolution of life is one of the most puzzling conundrums in modern physics and cosmology.

How do we get from the smooth and uniform CMB to the formation of structure in the universe? The first calculations of the fluctuations that lead to the seeds of galaxies in the CMB were presented at the Nuffield Workshop on the very early universe in June-July 1982 at Cambridge University by Guth and So-Young Pi, Hawking, Starobinsky, James Bardeen, Michael Turner and Paul Steinhardt. In 1981, V. F. Mukanov and G. V. Chibisov had also suggested that those fluctuations could be important, using Starobinsky's model. The calculations suggested that small clumps of matter or temperature differences in the CMB would become the seeds of galaxies as the universe expanded. The claim was that these little clumps would originate in the submicroscopic realm of quantum mechanics, and would be stretched enormously by the inflation of spacetime.

Quantum fields fluctuate in the vacuum from one point of space to another. Due to Heisenberg's uncertainty principle, inflation cannot erase these fluctuations at the quantum level because they are continually produced, even as the universe expands. Eventually, through the process of spacetime expansion, these fluctuations cross into the visible horizon. They become "frozen" into spacetime and imprint themselves on the surface of last scattering. As for classical matter and energy—quarks, electrons, and photons—it was present in the original cosmic soup. But according to inflation theory, inflation vacuumed it away, emptying out the universe by diluting all the matter to almost nothing. After inflation, matter and energy had to be reinstated to produce our present universe. This is achieved by what is called the "reheating" of the universe.

Returning to the analogy of the ball, we see the inflaton particle rolling along the plateau, and then suddenly descending to the bottom of the valley at its minimum potential energy, where it rocks back and forth for a while, and then settles and "decays" into other particles. This decay process reheats the universe from the very low temperature brought on by inflation, and matter in the form of quarks and electrons now appears in the universe again.

PROBLEMS WITH AND VERSIONS OF INFLATION

By the late 1980s there was extensive research to develop and refine different inflationary models. However, no matter what the details of the inflationary model, theoretical problems still exist with inflation.

One problem is that the particle physics interpretation of inflation is based on the particle or field called the "inflaton," which causes the inflation. This purported particle has a very tiny mass. As such it does not fit in well with all the other particles in the standard model of particle physics, and its incredibly small mass makes it virtually undetectable.

The second problem is that in order to produce exponential inflation on the cosmic scale, it is necessary to invoke a huge vacuum energy associated with the cosmological constant. However, the vacuum energy as measured in the present universe is critically balanced at a tiny value of about 10^{-29} to 10^{-30} grams per cubic centimeter. This is the critical density that produces the spatially flat Euclidean universe we inhabit today. The vacuum energy required in the very early universe to produce the violent, exponential expansion of the cosmic scale is about 10^{70} times the observed critical density of the vacuum energy in the present universe. How did that ever come about?

A third major problem faced by inflation is that the potential energy associated with the inflaton must be almost constant for a long time, that is, long relative to the very short periods of time in the early universe, in order to produce enough inflation to solve the horizon and flatness problems. Achieving this "flat" potential energy constitutes an unnatural fine-tuning of the physics, with the associated cancelling of large numbers to many decimal places. It is the Achilles heel of inflation theory.

Finally, to return to Hawking and Collins's objection, an overriding problem with inflation theory is the extremely low probability that inflation could actually have happened in the early universe. The initial assumptions that have to be made to get the inflation going are just too improbable.

In Guth's 1981 paper, he proposed a model of inflation based on a Grand Unified Theory (GUT). This was a popular endeavor at the time, as theorists tried to unify all the known forces of nature aside from gravity: the strong, weak, and electromagnetic forces. The forces were theorized to unite at a very high energy of about 10^{25} electron volts, or ten trillion trillion electron volts, and a huge temperature of about 10^{28} degrees Kelvin. Inflation was supposed to begin at this enormous energy level and cease shortly afterward. But of course we can never verify this idea directly, because we can never achieve such high energies and temperatures in our accelerators or laboratories.

In a subsequent paper, Guth showed that this original model could not succeed because it would produce a huge inhomogeneity in the present universe, in direct contradiction to the relative smoothness of the universe that we see in the CMB data today. To attempt to deal with this problem, Paul Steinhardt and his student Andreas Albrecht at the University of Pennsylvania suggested a different version of inflation in 1982 called "new inflation," which appeared to get rid of the problem of the large unobserved nonuniformity in Guth's model. This, however, also ran into difficulties, which led the Russian-American physicist Andrei Linde of Stanford University in 1983

FIGURE 8: Eternal inflation creates bubble universes through pockets of inflation.

to suggest what has been called "chaotic" or "self-reproducing" or "eternal" inflation.* In Linde's model, many small pockets of inflation are created in the very early universe; he theorized that out of a huge number of such small inflating universes, at least one would fit the bill and solve the initial value problems of our universe.

This version of inflation overcame some of the problems in Guth's original model, as well as deficiencies in the Steinhardt-Albrecht model.

BUBBLES AND LANDSCAPES

Many creative and imaginative minds have been pondering what the universe looked like and acted like in its earliest stages. "Bubbles" and "landscapes" arise from ideas about the inflationary universe. Linde's model of a self-reproducing universe can yield the finely tuned, flat, inflaton potential necessary for inflation. He pictured many universes as spherical bubbles, all of which contain an inflaton mechanism for inflation. The idea is that one of these bubble universes has a flat enough inflaton potential energy to produce exponential inflation without any extreme fine-tuning of the initial conditions of the universe. It would also be able to explain the smoothness of spacetime that is seen in the CMB.

In Linde's model, the bubbly inflation continues forever, eternally, beyond our universe's visible horizon. It is still happening today. This concept gets very close to an anthropic or a multiverse picture of a large or even infinite number of possible universes.† While the eternal inflation idea has support from Guth, Linde, and Vilenkin—who helped pioneer the idea of the anthropic principle—others criticize it. The eternal universe or multiverse idea has been criticized for losing predictive power because of its possibly infinite number of solutions.

Bubbles and landscapes have been around for a while. Back in 1977, Frank DeLuccia of Princeton's Institute for Advanced Study, and the well-known Harvard theoretical physicist Sidney Coleman published a paper in which they postulated that quantum vacuum fluctuations caused bubbles to appear in the vacuum potential energy landscape, and these bubbles could

---

* Steinhardt and independently, Alexander Vilenkin, another Russian-American physicist, first proposed in 1982 the "eternal" bubbling universe, and Vilenkin showed that eternal inflation was a generic feature of inflationary models.

† We recall that the anthropic principle states that certain constants and laws of physics are the way they are because otherwise life would not exist in our universe.

eventually tunnel their way through the cosmic landscape to become real universes.

In the late 1970s, the mathematical idea of bubble formation in an expanding universe was formulated with no reference to what we now call inflation—it arose by analogy to the behavior of liquids. It is possible to cool a liquid to a temperature significantly lower than its freezing point without it becoming solid. Such a liquid is called "super-cooled." Assuming the liquid is water, if you then introduce a small piece of ordinary ice into it, the water will rapidly crystallize around it. On rare occasions a large ice crystal will spontaneously form within super-cooled water and will grow enormously, making the super-cooled water freeze. This phenomenon is called "bubble nucleation" because the forming ice crystal can be treated as an expanding bubble.

Coleman and DeLuccia imagined that in the vacuum environment, the surface tension of the bubbles would exceed the force expanding them, squeezing them so much that they would disappear. But now and then, a bubble would become large enough to grow and survive for a while. Coleman and DeLuccia calculated the rate at which bubbles could be created in an expanding universe and found that the rate was very, very small but not quite zero.

The inflating vacuum is self-reproducing in the sense that space is constantly being created. A competition arises between the rate of growth of bubbles and the inflating expansion of space. At the quantum level in this expanding spacetime, occasionally a bubble manages to "tunnel" through an adjacent high mountain to enter a neighboring valley of lower altitude, or energy. This metaphoric process can repeat itself endlessly, producing large or infinite numbers of pockets of potentially inflating vacuum energy in an ever-growing energy landscape of valleys and mountain peaks. Another Stanford physicist, Leonard Susskind, has compared this far-out cosmological landscape, which is also the landscape of string theory, to a Darwinian evolutionary process with inflating pocket universes being created when the cosmological constant is large enough: Only the fittest pocket universes survive. Susskind and his collaborators claim that string theory—specifically M-theory, the biggest string theory of all—can explain the exponentially large number of valleys and mountain peaks in this landscape.

Are such bizarre scenarios merely wild, poetic speculation that can never be validated by experiment or observation? The number of critics of the bubbles and landscapes grows daily. Perhaps this excursion into eternal inflation's fantastical early universe cosmology will make the notion of a faster speed of light at the birth of time seem almost conservative in comparison.

ALTERNATIVE TO INFLATION:
THE VARYING SPEED OF LIGHT COSMOLOGY

In 1992, I proposed an alternative to standard inflation theory in which, instead of the cosmic scale of the universe inflating, the speed of light was very large in the early universe milliseconds after the big bang. I do mean very large—about 10^{29} (or 100,000 trillion trillion) times the currently measured speed of light! This did away with the initial value problems of the big bang model—the horizon, homogeneity, and flatness problems—just as effectively as inflation. A much faster speed of light in the infant universe solved the horizon problem and therefore explained the overall smoothness of the temperature of the CMB radiation, because light now traveled extremely quickly between all parts of the expanding but not inflating universe. A variable speed of light (VSL) also predicted that if the speed of light were initially huge, and then quickly slowed to become the present speed of light, it flattened out the universe. This avoids the extreme fine-tuning of parameters necessary in the standard big bang model.

The core of the VSL idea is to modify Einstein's relativity theory to allow for a violation of Lorentz invariance, the symmetry of special relativity. This would allow for a varying speed of light. The basic idea of Lorentz invariance, as we recall from Chapter 2, is that the laws of physics remain the same when you transfer from one uniformly moving inertial frame to another. In Einstein's special relativity, the speed of light is an absolute constant with respect to any moving observer. Thus, to make the speed of light vary, we have to postulate that the basic symmetry of special relativity, Lorentz invariance, is spontaneously broken.[1] When you do this, the laws of physics *do not remain the same* as you move from one coordinate frame of reference to another. Einstein's theory of gravity was also modified by me to allow the speed of light to change in the early universe. The upshot of all this is that the speed of light is only a constant in one particular frame of reference at rest, contradicting the basic postulate of Einstein's relativity theory. This modification of Einstein's gravity theory led to a consistent framework for the idea of a VSL in the early universe.[2]

The sudden change in the speed of light in the early universe was in the form of a discontinuous phase transition, like super-cooled water suddenly turning into ice. It is this phase transition, in fact, that caused the spontaneous violation of Lorentz symmetry. The dramatic change in the speed of light

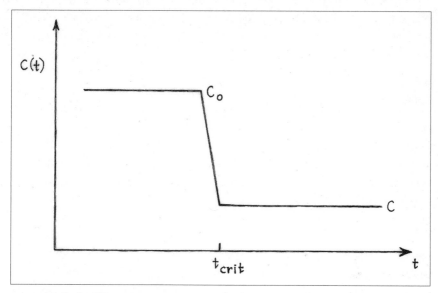

FIGURE 9: **In the VSL model, the very large speed of light in the early universe c_0 undergoes a phase transition at a critical time $t = t_{crit}$ and the speed of light becomes its current measured value.**

in the early universe can be compared to light passing through a refracting medium such as water or air. When the light rays strike the surface of water, they slow down and bend in direction. The speed of light is actually changing as light moves into the water.

Of course, in the early universe we do not have a medium such as water or air through which the speed of light can move and change. However, we do talk about physical quantities undergoing a phase transition at some critical temperature. When we reach a critical temperature in the very early universe, the speed of light, which may be a function of the temperature, suddenly changes discontinuously from its very high value to its currently observed value. This happens as the universe is expanding from $t = 0$ in the first fractions of seconds in the universe, during the very high initial temperature and energy conditions. I conceived of the speed of light as being enormous in the early universe a tiny fraction of a second after the big bang and then decreasing rapidly to its currently observed value of 300,000 kilometers per second.

In other words, the phase transition in the speed of light creates a scenario equivalent to inflation. The superluminal speed of light would causally connect vastly separated parts of spacetime beyond the visible horizon. Instead of spacetime inflating to iron out the wrinkles on its surface and creating a larger horizon in the early universe, the faster speed of light during the same time period would accomplish the same things by removing the horizon altogether, so that the entire universe could be seen. Recall that in inflation theory there is also a superluminal phenomenon as the universe inflates. Inflation itself proceeds at a speed faster than the measured speed of light. However, this superluminal behavior happens beyond the horizon of the early universe, so it would not be observed as a violation of special relativity. In VSL, on the other hand, the changing speed of light blatantly violates special relativity within the visible horizon.

I submitted my VSL paper, titled "Superluminary universe: a possible solution to the initial value problem in cosmology," to the *Physical Review D*. When the referee reports came back, a couple of referees were outraged by the very idea that I would change relativity theory in order to provide an alternative to inflation, and the editors rejected my paper. I next submitted the paper to the *International Journal of Modern Physics D* in Europe, where the referee and editor accepted it as written, and it appeared in 1993.

My graduate students at the University of Toronto were supportive of my pursuing the idea of a variable speed of light cosmology, and I benefited a great deal from their criticisms and suggestions. Virtually no one else in the physics community took any notice of my VSL work. The paper was ignored, and I turned to other projects.

Five years later, in 1998, I was scrolling through the abstracts of papers in the Los Alamos laboratory electronic archives when I noticed a paper entitled "Variable speed of light as a solution to initial value problems in cosmology" by Andreas Albrecht and João Magueijo. When I read the paper, I discovered that their ideas were strikingly similar to mine and yet there was no reference to my paper. The basic formulation of the VSL model in their paper was the same as I had proposed, although I had attempted to provide a more fundamental modification of gravity theory than they did. I noted that their paper had been accepted for publication by the conservative *Physical Review D*, which had rejected my paper six years earlier.

After several exchanges of e-mails between myself and Magueijo, it became apparent that Albrecht and Magueijo truly had not known about my paper; they had arrived at the VSL idea independently some six years later. Magueijo and I have since become friends. We have spent hours discussing our research at the Perimeter Institute, where he was on leave for two years from Imperial College, London. We have recently published a paper together on VSL, commenting on a critical review of the VSL ideas by the cosmologist George Ellis. Since the publication of the Albrecht and Magueijo paper and a related paper by John Barrow at Cambridge, the number of citations of my original paper has risen dramatically, and now it is considered a "famous" paper in the cosmology literature.

Magueijo published a popular science book called *Faster than the Speed of Light* in 2003. In it, he tells how he and Albrecht developed the theory of VSL but worried for two years about revealing the heretical theory to the physics community; how they went through a very tough time getting *Physical Review D* to publish it; and how taken aback they were to discover that I had already published a very similar idea six years previously. As João puts it in his book, "Imagine my shock when I discovered that another physicist had been there before us. As we landed, a flag already flew over the Moon."

BIMETRIC GRAVITY AND VSL

One beneficial effect of this scuffle over intellectual property was realizing that since others were now interested in the subject of VSL, it might be the right time to delve further into it. In 1998, rather quickly in collaboration with Michael Clayton, a former graduate student, I published a further development of VSL based on a bimetric geometry.

In contrast to Einstein's general relativity, which only has one metric, or metric tensor, of spacetime, a bimetric theory has two, and they are usually distinguished by one coupling to matter and the other determining the purely gravitational properties of spacetime. A gravity theory based on a bimetric geometry was first proposed by Nathan Rosen, who was Einstein's assistant at Princeton in the 1930s. Experimental data proved that this theory was not physically viable. A later version was proposed in the 1980s by Jacob Bekenstein for the purpose of solving the problem of gravitational lensing—the bending of light by the curvature of spacetime—in an alternative gravity model proposed by Mordehai Milgrom called Modified Newtonian Dynamics (MOND). Clayton and I found scenarios in which the speeds of light and gravity waves were not identical, and in which one speed could vary while the other remained constant.

In Einstein's theory of gravity, the speed of gravitational waves and the speed of light are identical. In fact, within the solar system, the speed of gravity waves—or the speed at which the pull of gravity is communicated between one massive body and another—has not been determined to better than 5 to 10 percent accuracy. Whatever the speed of gravitational waves, we infer from solar system astronomy that it is close to the speed of light.

One frame of reference in our bimetric theory is the VSL frame, while the other is the varying speed of gravitational waves (VSGW) frame. In the latter frame, the speed of light is kept constant, and the speed of gravitational waves varies in time. This means that if the speed of light is constant in the very early universe, then the speed of gravitational waves becomes very small, allowing for the universe to inflate. In the converse situation, when the speed of gravitational waves is kept constant in the VSL frame, then the speed of light can be very large in the early universe. This eliminates the problem of one part of the universe being unable to communicate with another, thus solving the horizon problem.*

In the gravitational waves frame, in which the speed of light is kept constant, a solution of the bimetric gravity field equations leads to an inflationary universe without the need for a fine-tuned inflaton potential energy. In other words, the bimetric gravity theory can predict either an inflationary universe or a varying speed of light universe, depending upon which frame one uses. The mechanism for producing an inflationary spacetime is the vanishingly small speed of gravitational waves in the first fractions of seconds after the big bang, which allows the universe to inflate enormously.

How do we distinguish between these two versions of the VSL theory, the one with the spontaneous violation of Lorentz invariance, which I published in 1993, and the bimetric gravity theory, published with Michael Clayton in 1998? I am still torn between favoring one version or the other.

VARYING CONSTANTS

The idea of varying constants is not new. In 1937 Paul Dirac, one of the discoverers of quantum mechanics, published a famous paper in *Nature*, "The cosmological constants," proposing that Newton's gravitational constant G varies in the early universe. He wrote in the paper:

> The fundamental constants of physics, such as c the velocity of light, $\hbar$ the Planck constant, e the charge and m_e the mass of the electron, and so on, provide us a set of absolute units of measurement of distance, time, mass, etc. There are, however, more of these constants than are necessary for this purpose, with the result that certain dimensionless numbers can be constructed from them.

---

* Similar ideas on a bimetric gravity theory were proposed independently by Ian Drummond of Cambridge University. Matt Visser, Stefano Liberati, and collaborators also published a development of VSL bimetric theory.

What Dirac was implying is that you do not need to keep all of these "constants" constant.

The problem is that as certain laws of physics developed over time, they required that particular physical quantities be called "constants." For hundreds of years before special relativity, the speed of light was not considered a constant, as in Newton's gravity theory and in later ether theories. So there was no built-in prejudice about varying this constant. Similarly, other "constants" of physics, like Newton's gravitational constant G, and Planck's constant $\hbar$ *were never absolutely constant* over the history of the universe. They only *appear to be constants* during certain epochs in the evolution of the universe. Yet we are left with this strange situation of having to speak of a "varying constant," which is an oxymoron. The problem is that the so-called constants like c and G are only constants within the context of a specific theory, such as Maxwell's equations of electromagnetism, Einstein's special relativity, and Einstein's theory of gravity. In these theories, the speed of light and the gravitational constant are postulated to be constant for all time. Given that this is true, then, we can of course set these constants equal to unity, thereby choosing a special set of units to measure distance, time, and mass.

There appears to be a visceral suspicion among physicists if one dares to change a constant. This was a major reason that the whole idea of a variable speed of light cosmology was subjected to criticism in the wake of the Albrecht and Magueijo paper. In 2001, back-and-forth volleys of rhetoric in e-mails and papers began among various physicists. A paper by Michael Duff, Lev Okun, and Gabriele Veneziano, titled "Trialogue on the number of fundamental constants," published in the *Journal of High Energy Physics* in 2002, epitomized this discussion. The three authors argued with each other in the paper about the meaning of varying the fundamental constants. Duff argued strongly that you could only meaningfully discuss varying constants in nature if they were dimensionless, like alpha, the fine-structure constant that measures the strength of the interactions between the electromagnetic field and electrically charged matter. Such fundamental constants as the speed of light and the gravitational constant, on the other hand, he claimed, have dimensions and cannot be changed. The speed of light has dimensions of distance divided by time. Duff declared that you could not even *consider* varying one of these constants.

In 2001, a team of Australian astronomers from the University of New South Wales led by John Webb, and the theorist John Barrow from Cambridge University (who three years before had published a version of VSL), published a paper in *Physical Review Letters* claiming that they had observed a change in the fine-structure constant alpha—it was increasing with time as the universe expanded. Thus if you extrapolated back in time to the beginning of the universe, they said, alpha would decrease in size.

Alpha is equal to the square of the electron charge divided by the product of Planck's constant and the speed of light. Therefore if c, the speed of light, were large in the early universe, and then decreased to its present measured value, one would expect alpha to be increasing as the universe expanded to its present-day size,* which the Australian astronomers had observed. However, other teams of astronomers have since analyzed similar data and found results that question the idea that alpha changes over time. Also, measurements of alpha taken in the unusual natural nuclear reactor in the Oklo mine in Gabon, West Africa, do not show alpha changing in time, either. However, because the epoch that these measurements correspond to in the history of the universe is different from that studied by the Australian team, the question of a changing alpha is still somewhat up in the air.†

One can ultimately resolve the issue of varying physical quantities by having two scales of speeds in a physical system. Thus in the bimetric gravity theory, two speeds naturally occur. The ratio of the speed of light to the speed of gravity waves is a dimensionless quantity, and this can be varied without worrying about changing physical units. Indeed, when discussing the varying speed of light theory, we are required to bring in gravity, so it is natural to form the dimensionless quantity that expresses this ratio.

WHICH IS IT, INFLATION OR VSL?

What is the actual likelihood of either inflation or VSL happening in the very early universe? In a paper published in 2002, Stefan Hollands and Robert Wald from the University of Chicago argued that the probability of initiating an inflationary early universe is almost zero. They likened it to a blind god throwing darts at a dartboard. The probability of the dart hitting the bull's eye of a big enough inflation to create our universe is almost zero. It was to overcome this dartboard probability problem, uncomfortably reminiscent of the Hawking-Collins balancing pencil critique of the big bang model itself, that led Andrei Linde to promote his "eternal chaotic inflation" scenario:

---

* The values of electron charge and Planck's constant would remain fixed through time.

† New analyses of the quasar spectroscopic data by Webb and his collaborators appear to further support their claim of a varying alpha.

With enough universes in a multiverse landscape, surely inflation could occur in at least one of them!

Eternal inflation produces an infinite number of bubbles or pocket universes that may have different laws of physics. At present, there is no reliable way to measure the relative probability of getting pocket universes like ours versus different ones. The notions of landscape and multiverse only exacerbate the problem.

Given that the chances of inflation happening in the early universe are so slim—and the phenomenon could only come about by extreme fine-tuning—the sensible reader might ask: Is inflation actually an improvement over the original big bang model? Is the fine-tuning of inflation 80 percent better than the fine-tuning of the big bang model? Is it 40 percent better? Or perhaps only 10 percent better? When one considers the problems of inflation in this way, the original motivation for inflation as a solution to the initial conditions of the universe is somewhat diminished.

Every physical model of the universe is based on some basic postulate that most often can only be indirectly proved true. Even in Newton's gravity theory, one has to postulate that space is three-dimensional, and yet we are unable to explain *why* this is so. Similarly in VSL models, we cannot explain *why* the speed of light has a very large value in the early universe—this is simply a basic postulate required to invoke a VSL model.

Both my original VSL model and the later bimetric gravity model predict results that agree with the smoothness and homogeneity of the cosmic microwave background data. However, in both models it is necessary to modify Einstein's gravitational field equations, which is not necessary in standard inflationary models. Einstein's theory contains a possible solution for inflation, but not for a varying speed of light. In 1911, Einstein did suggest a variable speed of light gravity, but it did not survive his later general relativity theory because the symmetry of special relativity and general covariance ruled out the possibility of the speed of light changing.

WILL THE DATA DISCRIMINATE
BETWEEN INFLATION AND VSL?

Despite problems and criticisms over the years, inflation has become accepted as a realistic picture of what happened in the very early universe. Why is more attention not paid to VSL, which is an alternative to inflation, solves the horizon and flatness problems, and predicts a scale-invariant spectrum of quantum fluctuations in agreement with experimental data? The answer may be more sociological than scientific. Inflation has been around a dozen years longer than VSL, and it is often easier to support the incumbent and not bother looking very deeply into the qualifications of the challenger.

But will experimental data or observations ever be able to discriminate between inflation and VSL, or some other scenario that resolves the initial value problem in cosmology, and show which scenario is true to nature?[3] Because of the opaque surface of last scattering, which prevents us from observing physical events before about 400,000 years after the big bang, we *can never directly observe* whether the very early universe near t = 0 underwent an inflationary expansion, or whether instead the speed of light was very large for fractions of seconds as in VSL models. Therefore, one's belief in inflation or VSL becomes akin to belief, or not, in a particular dogma. However, a powerful *indirect* form of information penetrating the surface of last scattering could possibly differentiate between VSL and inflation: primordial gravity waves.

In contrast to light, or photons, the gravitational waves would have been able to push right through the opaque surface of last scattering, bringing us a picture of events in the early universe. This is because the graviton, the carrier particle of gravity, only interacts very weakly with other particles such as protons or photons. Ghostlike, gravity moves right through matter. Inflation and VSL models provide specific predictions about the spectrum of gravitational wave fluctuations produced in the very early universe—and they are different from each other. Thus the signatures of the fluctuations of gravitational waves, if and when they are detected, may actually tell us which explanation of the early universe is correct—or, perhaps, whether they are both false.

Gravitational waves could only be detected if they were produced by some cataclysmic event in the universe, such as the collision of two black holes or the big bang itself. Gravity is so weak that only such a violent event could produce a strong enough signal to be detected. This cataclysm would radiate ripples in spacetime that would have a minuscule amplitude. Yet when passing by an object one meter in length, an ancient gravity wave could move it by about 10^{-13} meters, or one-thousandth the size of a hydrogen atom! However, in practice, to detect gravitational waves, we need a displacement sensitivity of one ten-thousandth the size of an atomic nucleus.

Joseph Weber, a physicist at the University of Maryland, founded a research program to detect gravitational waves in the 1960s. He invented a simple device to detect gravitational wave motion. It consisted of a large, solid piece of metal containing electronics that could detect any vibration. Unfortunately, Weber bars are not sensitive enough to detect anything except extremely powerful gravitational waves above the considerable background noise of sound waves. In the 1960s, Weber claimed to have observed his bar signaling the detection of a gravitational wave. This caused much controversy, and eventually his claim was refuted. Sadly, in his lifetime Weber never received the credit he deserved for initiating what is now a considerable international enterprise to discover gravitational waves.

Since Weber's work in the 1960s, a new detecting device called a laser interferometer has been invented. It has separate masses, placed many hundreds of meters to several kilometers apart, that act like two ends of a Weber bar. The most sensitive of these devices is LIGO—the Laser Interferometer Gravitational Wave Observatory. The problem with ground-based detectors is that their sensitivity to detecting gravitational waves is limited by low frequency sounds such as seismic noise, passing trains and cars, and even waves crashing on a shoreline hundreds of miles away. LIGO attempts to correct for this by placing one of its three "arms" in Louisiana and the other two in Washington State. However, it would be much better to detect gravitational waves at low frequencies in space. One such system is LISA: the Laser Interferometer Space Antenna, a joint European Space Agency (ESA)-NASA project scheduled for launch in 2011. LISA will have test masses placed five million kilometers apart, in separate spacecraft, with lasers running between them. (LISA will still be affected by other noise sources, such as solar wind and cosmic rays.) Because of the size of the LISA project in space, it will be able to detect very low vibrations, thus increasing the possibility of finding gravitational waves.

If LIGO and LISA are not enough to do the job, some scientists have even proposed using the moon as a giant gravitational wave detector. The moon's shape would be distorted an infinitesimal amount by gravitational waves, making it act like a giant Weber bar.

So far, LIGO and its upgrades such as the Italian-French VIRGO have not detected gravitational waves. If they eventually succeed, or if LISA succeeds in space, this would be a fundamental discovery. It would provide one more proof of Einstein's gravitational theory, which predicts that gravitational waves exist. It could also tell us whether standard inflation, VSL, or another scenario is the correct description of the very early universe. That is, it could tell us which alternative is better at "fixing" the standard big bang model of cosmology—and thus becoming part of it.

Chapter 7

NEW COSMOLOGICAL DATA

The only cosmological data that Guth and other early inflationists had to tell them about the very early universe were the radio astronomy CMB data of Aldo Penzias and Robert Wilson of 1964. These data showed that the temperature of the CMB was remarkably uniform throughout the sky. Since 1989, our understanding of the early universe has advanced with exciting new data from NASA satellite missions and astronomy. Physicists are now grappling with all this new information, trying to understand how the existing standard model of cosmology—consisting of Einstein gravity, dark matter and dark energy, the big bang theory and inflation—can accommodate them. Or will that model have to be changed or even abandoned in light of the new data?

SNAPSHOTS OF THE INFANT UNIVERSE

The Cosmic Background Explorer (COBE) satellite mission, a NASA project led by John Mather and George Smoot, was launched in November 1989. The COBE satellite took a picture from space of the cosmic microwave background (CMB) radiation that Penzias and Wilson had discovered in 1964. That picture is rich and startling, a snapshot of the oldest visible part of the universe. This remnant of the early universe—variously called the "surface of last scattering," the "era of recombination," or the "decoupling"—surrounds us like a giant sphere in the sky. Fossilized in the heavens almost fourteen billion years ago, the CMB represents the afterglow of light left over from the big bang. As we recall, the whole universe is bathed in this light, which today we no longer see as visible light. The radiation wavelength has been stretched by the expansion of spacetime, until it now consists of microwave radiation. This, the oldest light in the universe, has been traveling outward in every direction with the expansion of the universe from the surface of last scattering. The pattern of this radiation across the sky encodes a wealth of details about the earliest shape, size and content of the universe, and its ultimate fate.

The earliest COBE satellite pictures from the early 1990s reveal a smooth and uniform temperature of approximately 2.7 degrees Kelvin across the whole cosmic sky—that static snow in a faulty television set. As we recall, because spacetime and light have been redshifted by more than a factor of 1,000 since the formation of the celestial fossil 400,000 years after the big bang, we can calculate the temperature of the young universe at that time to be about 3,000 degrees Kelvin. This temperature was low enough to allow free electrons and protons to unite to form hydrogen and helium.

Further observations by the COBE team in 1992 revealed that the snowy picture of the CMB is not, after all, completely uniform, but as was predicted by cosmologists in the early 1980s, it contains tiny mottled patches that disturb the uniformity of the background CMB radiation by only one part in about 100,000. Finding the tiny patches in the CMB was so exhilarating when the team announced it that Stephen Hawking called it "the greatest discovery of the century, if not of all time."[1] As cosmologists had suggested, the tiny patches or fluctuations in the CMB are actually the seeds for the growth of galaxies and clusters of galaxies as the universe expands.*

After COBE came an equally amazing NASA endeavor, the Wilkinson Microwave Anisotropy Probe (WMAP), whose purpose was to study the tiny fluctuations that COBE had found. The WMAP mission sent back pictures and data that were analyzed by astronomers, resulting in the sharpest-ever image of the infant universe. WMAP's speckled portrait of the CMB confirmed the predicted temperature differences of just one-hundred-thousandth of a degree between patches in the sky. This mottled turquoise oval picture of the CMB's temperature, with its denser, hotter spots of yellow and orange, has become almost as famous an icon of our time as the photos of Earth from space brought back by the Apollo mission.

Many groups of astrophysicists spend every day programming large codes on computers in different laboratories in Europe and America trying to produce a "washed" version of the CMB data, in order to ensure that the data are reliable. When we observe the surface of last scattering, we have to remove what is called "foreground contamination." This is the contamination

---

* Mather and Smoot were awarded the Nobel Prize in 2006 for leading the COBE mission.

FIGURE 10: The **WMAP** picture of the cosmic microwave background (CMB), about **400,000** years after the beginning of the universe, depicts the temperature anisotropy as small splotches. Source: WMAP Science Team NASA.

in the data caused by galaxies and voids, the structure in our late-time universe that intervenes between us and the surface of last scattering. Getting an honest washed picture of the cosmic microwave background is very tricky. It depends on the model being used to describe the inhomogeneous regime in the late-time universe that is interfering with our observations of the early universe.

INTERPRETING THE NEW COSMOLOGICAL PICTURES

In what has become the new standard model of cosmology, the theoretical mechanism for generating the seeds that form galaxies is inflation and its associated particle, the inflaton. During the period of inflation, fractions of seconds after the big bang, the inflaton field experienced quantum fluctuations that were stretched exponentially in size, and crossed over the visible horizon to become "frozen" in spacetime. After inflation ceased, the universe continued expanding more slowly, and the visible universe caught up with the frozen fluctuations, which became imprinted on the CMB surface of last scattering. Finally the fluctuations grew through gravitational attraction to become galaxies.*

---

* The alternative mechanism to inflation, the Varying Speed of Light cosmology (VSL), suggests that the fluctuations were born out of spacetime itself during the period of radiation domination.

Let us look more closely at the early universe scenario in the standard model. Shortly after the inflationary period, the universe was dominated by hot radiation. This went on for about 400,000 years, until radiation and matter decoupled at the surface of last scattering. During this period before decoupling, matter was in the form of a plasma of ionized hydrogen—free protons and electrons—coupled to photons. In the standard model of particle physics, the protons are composed of bound quarks. We know from nuclear physics and particle physics that at very high temperatures, electrons and protons interact in such a way that photons cannot stream freely—they travel only a short distance before being scattered or absorbed.

This hot radiation plasma epoch after inflation—the cosmic soup—is crucial for understanding how galaxies grew. The protons (or their component quarks) in the plasma are tightly coupled to the photons, since the photons are the carriers of electrostatic attraction. The plasma can clump by gravitational collapse and begin to form larger and larger objects. It also possesses photon pressure, which counteracts the pull of gravity. Eventually, however, the gravitational force begins to dominate and more clumps form in the plasma, become larger, and then rebound due to the repulsive photon pressure. This expansion and contraction creates waves that travel at the

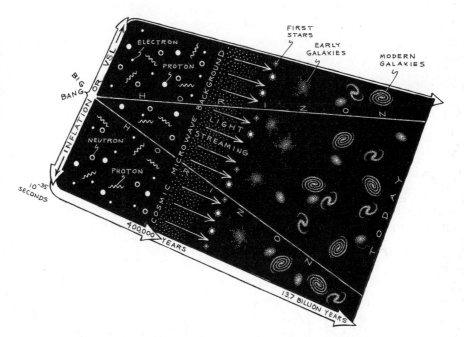

FIGURE 11: The universe timeline shows the evolution of the universe from the conventional big bang to the present.

speed of sound through the hot fluid, producing a kind of music in the early universe. We cannot hear the music anymore, but we see evidence for it in the CMB data as acoustical wave oscillations.

The clumps in the hot plasma can be seen in the CMB picture as temperature fluctuations. These are tiny compared to the overall temperature of the plasma, and therefore they can be treated mathematically as small perturbative fluctuations that yield approximate rather than exact solutions to the problems. It is important to understand *when* the gravitational pull in the plasma overcomes the photon pressure, because this determines how much the perturbation clumps can increase in size during the wave oscillations.

What favors the gravitational pull of the clumps over the pressure that diffuses them? In other words, what controls the growth or lack of growth of the clumps that eventually become galaxies and stars? Gravity begins to dominate when the gravitational in-fall time, or the time it takes for a clump to collapse, is less than the time it takes a sound wave to cross the clump. The balance between these two time periods is determined by the Jeans length or time, which was discovered by Sir James Jeans in the early twentieth century when he was studying the evolution of stars. Clumps larger than the Jeans length grow, while clumps smaller than the Jeans length do not. To produce galaxies, the inhomogenous clumps need to grow to the size of galaxies observed today, which can take more than a billion years.

TRACKING SOUND WAVES THROUGH THE ANGULAR POWER SPECTRUM

An important tool with which to probe the infant universe is the angular power spectrum, which has been determined with remarkable accuracy by the WMAP data.[2] "Angular" refers to the size of the angle in the cosmic sky that observers on Earth can see. "Power" in this context refers to the magnitude of the acoustical waves traveling through the cosmic soup. It is these sound waves as measured by the WMAP satellite at the surface of last scattering that are depicted in the angular power spectrum.

In fact, the angular power spectrum is a series of acoustical oscillations, or peaks and troughs, made by the sound waves as they travel through the hot plasma. They end up as fossil waves captured by the snapshot of the CMB. In the graph in Figure 12, the vertical axis is the strength of the acoustical power spectrum—that is, how big the waves are—while the horizontal axis represents the size of the angle within the CMB that we are observing. The angular power spectrum data show a large initial peak, which can be seen

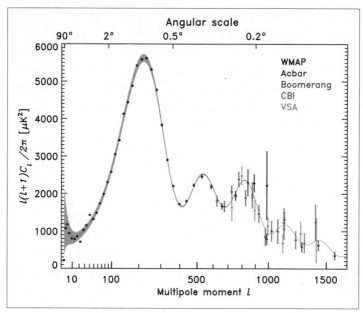

Angular scale

FIGURE 12: **The Wilkinson Microwave Anisotropy Probe (WMAP) angular acoustical power spectrum.** Source: WMAP Science Team, NASA.

when we observe a large enough angle of the sky to contain it. After that first major peak, we see pronounced second and third peaks in smaller angles of the sky, and thereafter a series of smaller ripples. What we are observing is the emergence of the photons from the surface of last scattering as the shapes of the musical sound waves in the early universe, formed almost fourteen billion years ago.

The WMAP data for the angular power spectrum can be augmented by more recent data collected by high-altitude balloons. In 2005, an international team of experimentalists published the results of an analysis of the balloon-borne experiments called "Boomerang." A balloon sent up high above the Earth's atmosphere recorded the cosmic microwave background, as the WMAP and COBE satellite missions had done. The balloon data indicate more strongly than the WMAP data the presence of a third peak in the power spectrum.

What do the sizes of these peaks, and their positions in the graph, tell us about the universe at 400,000 years of age? Remarkably, if one adopts the standard model, the position of the first peak in the WMAP data actually tells us the shape of the universe. It tells us that the geometry of space-time is spatially flat. This is because the position of the peak on the graph is determined by how curved space is. The WMAP angular power spectrum actually tells us that the universe is spatially flat to an accuracy within a few percent. This answers the long-standing question in cosmology as to whether the universe is flat, open, or closed. The three-dimensional universe that we perceive today has the structure of a three-dimensional Euclidean geometry, in which lines remain parallel to infinity and do not cross one another. If the universe continues to expand with this flat spatial geometry, then it will be infinite in extent, and the matter in the universe eventually will dilute to nothing.[3] However, that the universe turns out to be flat was not a complete surprise. Both the simplest inflationary models and VSL predict a spatially flat universe.

As well as telling us the shape of the universe, the angular power spectrum also gives us information about the matter and energy content of the universe. If we attempt to adopt a model in which there is no cold dark matter and no dark energy, then the hot plasma before recombination consists only of visible baryons and photons. If we use only Einstein's gravity theory to analyze the data, we find that the position of the first peak does not agree with the data, and the second and particularly the third peaks will be erased. The reason is that the strength of the pull of gravity is significantly reduced when one uses Einstein's gravity theory and only visible baryons and photons. Therefore the CMB data tell us that we must either include significant amounts of dark matter and dark energy when keeping Einstein's theory, or we must modify Einstein's gravity theory to exclude dark matter and provide an explanation for dark energy. Achieving a fit to the acoustical power spectrum data without dark matter is a major goal of any modified gravity theory.

THE MATTER POWER SPECTRUM AND THE MATTER CONTENT OF THE UNIVERSE

A second kind of power spectrum, the matter power spectrum, tells us more about the content and distribution of matter and energy in the universe. These data come not only from the CMB, but also from counting the galaxies in the sky. It is necessary to know how much matter is in the universe, because this basic information determines the large-scale structure of the universe. It is matter and energy that warp spacetime, and therefore determine the shape of the universe. In practice, cosmologists use this matter power spectrum to estimate the amount of dark matter and dark energy needed in the standard model to describe the universe using Einstein's theory of gravity. The graph of the data for the Sloan Digital Sky Survey (SDSS) matter power spectrum will prove to be very important in differentiating between the predictions of Einstein gravity with dark matter and the predictions of modified gravity (MOG), with its stronger gravity and no dark matter. Chapter 13 will discuss these differences—and this important test of MOG versus dark matter—in detail.

The matter power spectrum tells us the statistical distribution of matter and energy in the universe, and from that we can figure out the actual amounts. Think about the difference between a homogeneous and an inhomogeneous distribution of matter in space. In a homogeneous or uniform distribution, if we have a small clump of matter, then other similar clumps will be uniformly distributed throughout the entire space. In an inhomogeneous distribution, on the other hand, the clumps of matter appear to us in a nonuniform way. If we fly in a helicopter over New York City, we expect to see more people directly below us in Manhattan than we do when looking out toward Long Island, because more people live in cities than in outlying areas. If we construct a power spectrum of the distribution of people in a country, we statistically expect to find more people in cities than in rural areas. Thus when we talk about the matter power spectrum, the word "power" refers to the numbers and statistical distribution of objects, in this case not acoustical waves but galaxies. In a matter power spectrum of the early universe, there will be many clumps in areas where galaxies are forming, and very few in areas that will evolve into cosmic voids.

One way of picturing the whole sky, and viewing the galaxies or the visible matter in the universe, is as a cosmic sphere rather than as a flat map. One can gather these data from Earth-based telescopes, the Hubble space telescope and satellite missions. During the past several years, a picture of this cosmic sphere has been put together by two main groups of observational astronomers. One project in New Mexico consists of a team of 14 institutions using a 2.5 meter telescope. Since 1998 these astronomers have compiled the Sloan Digital Sky Survey (SDSS). They have measured the distances to over two million galaxies, thereby mapping out a quarter of the sky in the northern hemisphere. Another large survey at the Anglo-Australian Observatory in New South Wales, the 2dF (2 degree field) project, has mapped out the cosmic sphere for the southern hemisphere. The result of all these observations is a map of the statistical distribution of several million galaxies. Analyzing this matter power spectrum can tell us not only the present amount and distribution of matter in the universe, but also the matter content of the early universe as far back in time as we can count dim galaxies. The matter power spectrum in effect follows the history of the clumps in the CMB from recombination through star and galaxy formation until the present day.

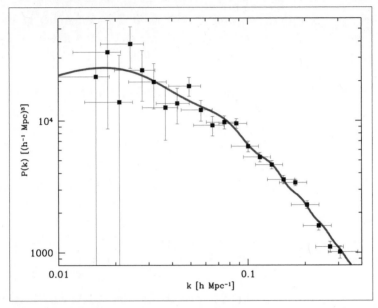

FIGURE 13: **The Sloan Digital Sky Survey (SDSS) matter power spectrum.**
Source: Max Tegmark/SDSS Collaboration.

There has been a recent flood of cosmological measurements for the CMB and measurements of the distributions of stars and galaxies. Figure 13 shows the power spectrum, or the distribution of matter, versus the wave length k, where k is the inverse of the distance scale that we are looking at on the cosmic sphere. The cube of this distance gives us the volume for a particular part of the cosmic sphere. As k goes to the left toward zero on the horizontal axis of the graph, the size of the length scales observed across the sky, or its volume, increases until we eventually obtain the volume of the whole visible universe. Conversely, when k becomes large toward the right side of the horizontal axis, the length scale across the sky and its volume decrease, until we only have shorter length scales representing the distances between clusters of galaxies and eventually galaxies themselves. The galaxy distribution data are represented by dots with error bars, and the dark curve is the best fit to the data predicted by the standard model of cosmology, using what is called the "Lambda CDM" model, where Lambda is the cosmological constant and CDM stands for cold dark matter.

According to the standard model, during inflation—which occurred *before* the data shown on this graph—spacetime expanded exponentially like a balloon, which made the perturbative fluctuations in the cosmic soup stretch as well. These primordial fluctuations then grew in size in such a way that the resulting power spectrum looks the same at all scales. The fact

FIGURE 14: **Fractal cauliflower, like the matter power spectrum from the early universe, looks the same no matter what the distance scale.** Source: Photo by John Walker.

that the curve (not shown in the graph) is a continually rising straight line means that the distribution of fluctuations is scale-invariant for this part of the graph.

This so-called scale-invariant spectrum is one of the chief predictions of inflationary models, and has thus been verified through the WMAP data.* Scale invariance in general means that the shape of any distribution of objects remains the same if we increase or decrease the size of the space in which they are distributed. A good example of scale invariance is fractal patterns. Study the so-called fractal cauliflower in Figure 14, which exhibits a scale-invariant repetition of the same shapes and patterns regardless of the level of detail one chooses to look at.

It is the same with the CMB part of the matter power spectrum: However one views the matter distribution of fluctuations, the pattern of distribution will remain the same.[4]

We see in Figure 13 that the data points representing primordial, protogalaxy distributions do not continue to increase indefinitely, but they turn over, producing a bump in the graph and then a continually decreasing curve all the way down to large values of k or small volumes of the cosmic sky.

---

* The original prediction of the scale invariance of the spectrum in the early universe was made by two astrophysicists, Edward Harrison and Yakov Zeldovich, before inflationary models were published.

At these smaller scales, we begin to see the power spectrum distribution of clusters of galaxies and eventually galaxies. The bump and the subsequent decrease of the curve are due to the growth of fluctuations that are clumping to form stars, galaxies, and clusters of galaxies.

How does the matter power spectrum help us to understand the constituents of matter in the universe? According to the standard model, at least 90 percent of the matter in the universe is "dark," including in the early universe. Since dark matter does not emit radiation in the form of photons, how can we measure the amount of it if we cannot see it? In estimating the amount of dark matter, cosmologists developing the standard model have to make some rather strong assumptions about their observations. For one thing, they assume that galaxies have dark matter halos around them that trace the visible light of galaxies. By knowing the number of galaxies, cosmologists then estimate the amount of dark matter in the universe. Of course, they can't give an absolute estimate, but they assume that the shapes of the matter power spectra for the visible clusters of galaxies and galaxies are always approximately the same. Therefore cosmologists expect that the shapes of the dark matter power spectra are also the same for different types of galaxies, for they assume that dark matter is always abundantly present when visible matter exists. However, astronomical data do not always confirm this—a situation that has triggered heated controversy.

PULLING IT ALL TOGETHER

In the 1960s, Fred Hoyle and his collaborators worked out remarkably accurate nuclear physics calculations that told them the density—and therefore the actual amount—of baryons in the early universe during the period of nuclear synthesis. This was between one second and 100 seconds after $t = 0$, when the temperature of the cosmic soup was about one billion degrees Kelvin, and isotopes and elements heavier than simple hydrogen, such as deuterium, helium and lithium, were formed.[5] Today, this period in the universe's early history would appear at a redshift of approximately ten billion, but we cannot "see" it because it occurred before the surface of last scattering; it is behind the curtain of the CMB. Because baryons are conserved throughout the life of the universe, the number that Hoyle came up with is the same number of baryons that exist in our present universe.

A major task of the modern cosmologist is to develop a model in which both the matter power spectrum and the angular power spectrum can be fitted to Hoyle's and his collaborators' calculations of the amount of baryons in the universe. If a particular cosmological model fits the data successfully, then it may well be describing accurately the large-scale structure of the universe. As it turns out, the standard concordance model, or Lambda CDM model, in which the WMAP data are analyzed according to calculations based on a computer code called CMBFAST, concludes that the universe is composed of about 4 percent baryons (quarks in the form of protons and neutrons), 24 percent dark matter, and 72 percent dark energy. This choice of matter and energy content fits the matter power spectrum data and the angular power spectrum data remarkably well, given the spatially flat geometry of the universe. I emphasize that these numbers are necessary when one attempts to fit the standard model, based on Einstein gravity, to the data. A different model might produce different numbers, or, as in MOG, a value of zero for exotic dark matter. This will be discussed in Chapter 13.

As for the dark energy, an expression coined by the University of Chicago cosmologist Michael Turner, it distinguishes itself from the approximately 30 percent of baryon matter and putative dark matter because it has to be uniformly distributed across the sky, whereas matter is not uniformly distributed. The dark energy cannot clump due to gravitational pull, because this would seriously contradict the model of how galaxies are formed in the early universe. In fact, if this amount of dark energy clumped in the early universe, it would stop galaxies from forming.

THE ACCELERATING UNIVERSE AND DARK ENERGY

Is there any evidence for dark energy? In 1998, two groups of astronomers at the Lawrence Berkeley National Laboratory in California and the Australian National University in Canberra were studying the light coming from dozens of type 1a supernovae. There are two basic types and several subtypes of supernovae. The one we are concerned with here, which is used as a celestial yardstick, type 1a, is a white dwarf star in a binary system. Gradually, it pulls in gas from its companion star until it becomes too heavy and collapses, triggering a thermonuclear explosion. It is assumed that all type 1a supernovae have the same intrinsic or absolute luminosity. Therefore they are as useful as the Cepheid variables in determining distances in the universe. Supernovae are quite rare. The last one that appeared in the neighborhood of our galaxy was in 1987, and the last to appear in the Milky Way itself was one in 1604 that Kepler observed. The supernovae being observed today are in other galaxies.

In 1998, the Californian and Australian astronomers, led by Saul Perlmutter and Brian Schmidt, respectively, discovered from luminosity distance measurements that the light coming from the supernovae was 50 percent dimmer than it "should have been," according to the estimates of their distances. The astronomers interpreted the dimness of the light as evidence that the supernovae were farther away than they should have been if the expansion of the universe was slowing down, which was what everyone assumed was happening. And so, they concluded that the expansion of the universe was actually *accelerating*—an astonishing result!

Critics questioned this conclusion, pointing out that interstellar dust between Earth and the supernovae could be causing the light to appear dimmer, or perhaps we were misunderstanding something about the evolutionary process of supernovae. However, the astronomers carefully rechecked all aspects of their observations, and again claimed that the interpretation of an accelerating universe was correct. They even suggested that the expansion of the universe had started speeding up some five to eight billion years ago, in spite of the fact that the influence of the matter should be exerting a big enough gravitational pull on the universe to slow it down.

Since, according to the standard model of cosmology, the dark energy is supposedly uniformly distributed in the universe, then surely the matter content of the universe was dominant at some time in the past, causing the expansion of the universe to *decelerate*. However, the matter density would have decreased as the universe expanded, and for some as-yet-unknown reason, the matter density must have fallen below the dark energy density in our relatively recent past, five to eight billion years ago, and as the dark energy began to dominate over the matter density, the expansion of the universe would have begun to accelerate.

There has been a great deal of speculation recently about the nature of dark energy. Nobody yet understands its origins, and we do not even understand how it evolved as the universe expanded from time $t = 0$. We cannot directly detect dark energy in the laboratory or through telescopes. One interpretation is that the dark energy is vacuum energy. Recall that in modern particle physics, the vacuum is not just "nothing," but it creates virtual particles like protons and antiprotons and electrons and positrons (antielectrons), which then annihilate one another to produce what we interpret as a vacuum. In other words, the creation and annihilation of protons, electrons, and other particles is perfectly balanced, resulting in the minimum energy of the universe, which we call the vacuum. Since the vacuum energy is constant, it is considered identical to Einstein's cosmological constant, which introduces a repulsive, antigravity force into the universe. In practice, the vacuum energy, dark energy, and the cosmological constant are synonymous.[6]

However, we recall that the cosmological constant has had a long and torturous history, and that it requires monstrous fine-tuning calculations to make it consistent with astrophysical observations of its value. So most physicists are troubled by interpreting the dark energy as Einstein's cosmological constant, and that is why many alternative interpretations of the dark energy have been cropping up. I will describe these alternatives, including my own, in Chapter 15.

On the other hand, it is extraordinary that when we figure out the amount of dark energy needed to make the universe flat and consistent with the Friedmann equations in the new standard cosmology, that amount also fits the acoustical peaks in the WMAP data. This means that the WMAP CMB data are in accord with the supernovae observations that show a speeding-up of the expansion of the universe. Indeed, most physicists and cosmologists agree that the *dark energy itself* appears to be causing the expansion of the universe to speed up.

I emphasize that we have been led to this new picture of the universe not through idle speculation but through our analysis of the hard observational facts revealed by the accurate WMAP data for the CMB. The data for the acceleration of the universe obtained from supernova observations are also combined with large-scale surveys of galaxies, distributions of clusters of galaxies, gravitational lensing and other observational data to form a picture of the amount of matter and energy that exist in the universe.

Yet we need to find a convincing theoretical explanation for this strange energy-matter content of the universe, in which so much is invisible. Why are most physicists convinced that only 4 percent of the universe consists of visible baryons? What is the nature of dark matter, the additional 24 percent of missing matter? Will we ever be able to observe it? Or are the effects seen in the data actually due to stronger gravity, as my MOG shows, and in fact there is no dark matter at all?

And what is the origin and what are the characteristics of the supposed 72 percent dark energy? Why is the universe accelerating its expansion now, in the epoch in which we are living? This question has given rise to the peculiar "coincidence problem": It seems preposterous that the period of acceleration should be happening now, in the epoch in which we are living, and not twelve billion years ago or eight billion years in the future. This places us in the dangerous position of considering the anti-Copernican possibility that we occupy a special place and time in the universe.

There is a race in progress now by many physicists, astronomers, and cosmologists, both on the theoretical front and the observational front, to answer these questions. Dark matter, dark energy, and the acceleration of the universe are three of the major problems facing modern physics and cosmology today.

The new standard model of cosmology, which I have presented in Chapters 3 to 7, is considered to be the correct description of the universe by a large consensus of astronomers and cosmologists. It is based on Einstein's general relativity, postulates the existence of dark matter in order to fit certain data, embraces inflation as well as a singular, big-bang beginning to a finite universe, contains black holes, and postulates a dark energy described by Einstein's cosmological constant. The equations of this standard model are built into the data analyses that astronomers and cosmologists use. Thus the standard theory significantly influences the process of data interpretation and creates a bias in its own favor.

We are now motivated to ask the question: What if we modify Einstein's theory—if we discover a new gravity theory—and in the process get rid of dark matter and explain the dark energy? This could possibly lead to a cosmology that fits all the observational data just as well as the new standard model but without having to press into service huge amounts of undetected and quite possibly *imaginary* dark matter.

PART 4
SEARCHING FOR A NEW GRAVITY THEORY

Chapter 8

STRINGS AND QUANTUM GRAVITY

S

o far I have mainly been describing classical gravity, which operates within the huge distance scales of the cosmos. Now I want to focus on gravity at distance scales as small as the unimaginably tiny distance of the Planck length of about 10^{-33} centimeters, which corresponds to the huge Planck energy of 10^{28} electron volts. Gravity is unimportant at the subatomic level until we reach this minuscule Planck length, when gravity again dominates all the other forces of nature.[1] Ideally, we seek a gravity theory that describes nature from the farthest reaches of the universe down to the smallest microscopic distances we can imagine. This is the goal of both string theory and the quantum gravity theories I shall discuss in this chapter.

For decades, there has been a consensus in the physics community that the two great pillars of modern physics—general relativity and quantum mechanics—must be combined. Both are very successful in providing consistent theoretical frameworks. Both have been verified by accurate experiments—quantum mechanics at the submicroscopic level and general relativity in astronomy. But combining the two into a larger theory has stumped brilliant minds for more than sixty years.

Many physicists agree that to unify general relativity and quantum mechanics, we have to "quantize" the gravitational field of Einstein's theory in the same way that we quantize the classical electromagnetic fields of James Clerk Maxwell's celebrated field equations. That is, just as we conceive of packets of quantum energy called photons for electromagnetism, we must conceive of gravitons for gravitation. And it is the gravitational waves, most physicists believe, that should be quantized in packets of energy. Even if we attempt to quantize spacetime itself, we still have to somehow identify the graviton as the quantum energy package of gravitational waves in order to reach a meaningful physical interpretation. Such a successful theory, if discovered, would be called the quantum theory of gravity.

The goal of a quantum gravity theory is less ambitious than the quest for a theory of everything, which would unify all the forces of nature. These goals have divided the theoretical physics community into two camps. The larger camp holds that string theory will lead to a theory of everything, including a successful quantum gravity theory. A much smaller camp maintains that a successful quantum gravity theory can be achieved without searching for a theory of everything, and indeed a quantum gravity theory would constitute a great success in its own right.

CAMP ONE: STRINGS AND BRANES

String theory had its origins in the 1960s in attempts to understand the strong interactions that bind quarks together inside protons and neutrons. Gravity was not considered part of the string picture at that time. Physicists including Gabriele Veneziano, Yoichiro Nambu, Holger Bech Nielsen, and Leonard Susskind proposed that the strong force holding the atomic nucleus together could be explained by depicting the particles within the nucleus as strings. However, this description was ultimately not successful. The string theory of strong forces was short-lived, but the mathematical idea of strings remained.[2]

By the early 1970s, many physicists felt that the failure of quantum gravity theories to produce finite, predictable results had created a serious impasse for physics. Some began searching for a new way to solve the problem of quantum gravity and at the same time unify gravity and the other forces of the standard model of particle physics. In 1974, John Schwarz and Jöel Scherk realized that of all the infinite number of vibrating strings that can exist in a string theory, one of them can describe the graviton, which must have zero mass with quantum spin 2. Abandoning the old idea of using string theory only to describe the strong nuclear force, Schwarz and Scherk claimed instead that they had discovered a unified theory of all the forces, including gravity, and that the size of a string is the almost infinitesimal Planck length.

When Scherk and Schwarz introduced their scenario with its tiny strings, including the stringy graviton, they hoped to overcome the problem of unphysical infinities in the calculations involving gravitons and particles of the standard model, reasoning that if a zero-dimensional point is replaced by a one-dimensional string with a size, this would remove the infinities associated with the point particles. Scherk and Schwarz were convinced that this was the only consistent way to combine gravity with quantum mechanics. They went as far as to say that if Newton had known about string theory, he would have predicted the existence of gravity!

An early model of vibrating strings was published by the Japanese physicist Yochiro Nambu and his collaborator T. Goto. The Russian-American physicist Alexander Polyakov, a professor at Princeton University, next developed a mathematical model of vibrating strings in the early 1980s, by generalizing the earlier work of Nambu and Goto. Polyakov's strings vibrated to produce tonal modes, much as the strings inside a piano vibrate when hit by a felt-covered hammer. The lowest tonal modes would then be identified with the observed particles in the standard model—an appealing idea, but one which unfortunately has not yet succeeded.

The first string theory "revolution," as proposed by Scherk and Schwarz, was not accepted initially as a viable research project by the community of theoretical physicists, in part because there were too many possible string theories. Secondly, some physicists were suspicious of the fact that string theory can only be consistently described in a higher-dimensional spacetime, that is, dimensions beyond the three dimensions of space and one of time that we observe and inhabit every day. Introducing higher dimensions hugely increases the number of physical degrees of freedom that can exist in string theory, as compared to theories that only reside in four dimensions. A significant increase in the number of degrees of freedom in a theory in turn leads to a huge increase in technical complexity. In 1984, a landmark paper by Schwarz and Michael Green showed that string theory could unify all the four forces and all of matter. Moreover, they resolved subtle conflicts between string theory and quantum physics.

For strings describing just bosons (particles with integer spin like photons, mesons, and gravitons, which mediate the forces between matter particles called "fermions"), the string theory has to reside in no fewer than twenty-six dimensions in order to be consistent with special relativity and to contain no unphysical negative energy vibrational modes. To include fermions (particles with half integer spin like electrons, protons, neutrons, and quarks) in string theory, we must also include supersymmetry in the string theory framework. Recall that supersymmetry theory is considered the simplest symmetry of particle physics; it states that for every boson particle there is a supersymmetric fermion partner, and for every fermion particle there exists a supersymmetric boson. In particle physics this symmetry doubles the number of particles that supposedly exist in the universe. Extensive searches for supersymmetric particles will soon be started at the Large Hadron Collider at CERN.

The supersymmetric string theory called "superstring theory" resides in ten-dimensional spacetime—nine spatial dimensions and one time dimension.* That is, the extra number of spatial dimensions beyond our perception of everyday reality is six. No one has ever seen these extra six dimensions of space. Of course, theorists are imaginative enough to hide this fact and explain the elusiveness of the extra dimensions by "compactifying" them, or rolling up these six dimensions into a tiny space that has the dimensions of Planck's length, of about 10^{-33} centimeters. This distance is so small that if you used it as a ruler to measure the diameter of an atom, you would have to lay it end to end 10^{25} times, or ten trillion trillion times. Needless to say, spaces of this size are not going to be detected in the near future. As of this writing, more than thirty years after the first publications on the subject, string theory has not produced *any* testable predictions, nor has anyone yet come up with a way to falsify it.

During these three decades, string theory has gone through various stages of development, somewhat reminiscent of French history in the late eighteenth and early nineteenth centuries. String theory has experienced two or three revolutions, each one triggering a new wave of enthusiasm among string theorists. String theory became a darling of the media as a stampede of physicists joined Green and Schwarz in seeking a "theory of everything," and enormous financial and manpower resources were applied to develop the theory. Perhaps one prominent feature of string theory is the use of the word "super," which enhances its attraction for physics students and the media alike. Only recently, with prominent critics of string theory gaining more attention, has string theory's glamor begun to fade.

The second string theory revolution occurred in 1985, when it was discovered that of the many possible string theories, only five could provide consistent and finite calculations. This was considered a triumph because it reduced a previously enormous number of superstring theories to a handful. A leader of the strings movement, Edward Witten, who is one of the most influential theorists in the physics community today, proposed that the theory

---

* Superstring theory in its later versions contains (after compactifying the extra six dimensions) a modified version of Einstein's general relativity, for the gravity theory has new fields such as spin 0 fields. Therefore superstrings should be considered a modified gravity theory.

could actually be an eleven-dimensional theory. Witten's idea was dubbed "M-theory." There are disputes about what the M stands for. Maybe it's the "master theory," for it contains the known five consistent string theories in ten dimensions? At the present time, there is not yet an actual M-theory. There have been proposals of how to interpret M-theory, but none has been particularly successful. Perhaps M stands for "Messiah," since waiting for M-theory to be formulated is a little like waiting patiently for the Messiah to appear on Earth. Since 1986, Andrew Strominger and others have demonstrated the mathematical richness of the solutions based on string theory. However, this richness was ignored until 2000, when it became clear that due to the infinite number of ways you can go from a ten- or eleven-dimensional theory to a four-dimensional spacetime theory, there is in fact not a handful after all, but indeed a very large number or possibly even an *infinite* number of string theory solutions. And these solutions translate to a very large or possibly infinite number of vacuum or minimal energy states.

It was not long in the history of string theory before someone hit upon the idea of extending the one-dimensional string to a higher-dimensional object called the "brane," shortened from "membrane." The idea of incorporating branes into string theory was first proposed by Joseph Polchinsky, a professor at the Kavli Institute for Theoretical Physics in Santa Barbara, California, in the 1990s, and it was championed by the English physicist Michael Duff while he was a professor at the University of Michigan.

However, many years before this, in the 1950s, Paul Dirac was the first to publish a paper on branes, when he proposed a model of the muon particle as an excited state of an electron. It was always puzzling to physicists why a particle should exist that looks like an electron but is much heavier and can decay into an electron plus a neutrino plus an antineutrino. Dirac had the idea that one could embed a two-dimensional brane into four-dimensional spacetime. By using the membrane idea, he sought to explain the muon as an excited state of an electron.

The modern introduction of branes cohabiting with strings made the whole subject far more complex. Soon string theorists were proposing that the branes had a charge, and the charge density on branes produced a flux when they moved. Branes are now an integral part of the theoretical machinery of the string theory paradigm.

THE STRING LANDSCAPE

The vacuum state, or the ground state of quantum mechanics, we recall, plays a fundamental role in quantum mechanics and quantum field theory. It is the state of lowest energy, in which no particles are present, and yet it is teeming with particles and antiparticles that are continually created and annihilated, producing a jittery fluctuation of the vacuum. In the standard model of particle physics, there is only one vacuum state, and this guarantees that one can make definite predictions that can be tested by experiments. But in string theory, the perhaps *infinite number of vacuum states* has created what is called the "landscape." This is not the landscape of garden designers but the metaphorical hills and valleys depicting the infinite number of vacuum states or possible maxima and minima of energies in the string theory solutions. The lowest point of the valleys represents the quantum mechanical ground state of zero energy.

In quantum field theory the vacuum has energy, and in gravity this energy is identified with the repulsive dark energy of Einstein's cosmological constant. The recent observation that the expansion of the universe is accelerating suggests that there must be some kind of dark, antigravity energy in the void of space that is fueling this acceleration. One of the most straightforward explanations for the acceleration of the universe uses Einstein's cosmological constant associated with the vacuum energy. The acceleration data determine an upper bound for the size of the cosmological constant or vacuum energy that is extremely small when compared to estimates of the constant based on quantum field theory calculations. The vacuum energy itself, determined by cosmological observations, is about 10^{-26} kilos per cubic meter—a tiny density indeed. The natural prediction of quantum field theory is that the cosmological constant is 10^{122} times bigger than what is permitted by cosmological data. This makes it the worst prediction in the history of science! This absurd prediction makes it hard for theorists to accept the cosmological constant as an explanation for the accelerating universe. The onus is on the practitioners of quantum field theory and superstring theory to resolve this serious conundrum. So far, the consensus in the physics community is that all attempts to use string theory to solve the problem have failed.

Can one of the valleys in the string landscape be a vacuum state that would lead to our universe and the tiny cosmological constant that has been observed? The answer now among prominent string theorists is that since there are many trillions, or even an infinite number, of vacuum states in the string landscape, then surely one of them represents the one observed in our universe. This, of course, amounts to a nonprediction. If you make the supply of solutions large enough in a theory, then you can always find an answer to every question.

STRINGS IN TANGLES: THE MULTIVERSE
AND THE ANTHROPIC PRINCIPLE

Impatience with string theory is finally being loosed in the physics community. Outspoken critics Peter Woit of Columbia University and my colleague Lee Smolin at the Perimeter Institute in Ontario have recently published books critical of string theory. They claim that the infamous and pockmarked string landscape with its infinite number of solutions is bringing on the demise of string theory. If string theory is to survive much longer, another revolution is needed. Some principle must be discovered that shrinks the landscape, curtailing the large or infinite number of string solutions or vacuua, so that string theory can finally predict something meaningful, such as: "This, *right here*, is the vacuum state that leads to the correct value of the cosmological constant and the standard model of particle physics."

Few ideas turn out to agree with nature, however interesting, beautiful, or unique they may be. This is the superb challenge of theoretical physics. Yet because of the large or even infinite—certainly out-of-control—number of solutions in superstring theory, no predictions have ever been made, let alone tested or falsified. A theory like Einstein's general relativity can be considered to have a sizable number of solutions—possibly an infinite number. However, the big difference, compared to string theory, is that when Einstein's equations are solved in terms of a set of measured constants (such as the mass of the sun and the gravitational constant), and a specific boundary condition is imposed on the solution (such as that the solution should become flat Minkowski spacetime at an infinite distance from the gravitational source), then the solution becomes unique. A specific prediction such as the motion of the planet Mercury can then be made and tested by observational data.

The problem with the huge and possibly infinite number of string solutions is that there is no way to control the landscape of solutions. There is no known principle in string theory that can tell us which particular solution or vacuum state corresponds to the one we observe in nature. We can't solve a set of equations that tells us which solution of string theory we should use to compare with observational data. This means that string theory and its related brane theory are mathematical models with no predictions that can be tested or falsified.[3] According to Woit and Smolin, this mathematical game playing, with no apparent grounding in reality, has created the current crisis in string theory.

In order to stave off the demise of string theory, prominent string theorists are now promoting a change in how we do science. Up until now, science has always tested theoretical hypotheses with factual data, to guarantee that the conceptual model fits nature. Now, in a desperate attempt to save string theory and to avoid the embarrassment of thirty years of wasted efforts, resources, and careers, some prominent string theorists tell us that we can't do science any more the way it has been done since Galileo, Kepler, and Newton. Instead, we are told that we shouldn't worry about testing theories or falsifying them, and we should simply accept that we cannot in the end determine fundamental constants like the cosmological constant! Indeed, the calculation of the cosmological constant becomes like the question, "Why are the planets in the particular orbits that they are around the sun?" This brings a disturbing randomness into physics.

Of course, to justify such a radical change of methodology—no less than the tossing out of the time-tested scientific method—we must have a principle. And luckily for the string theorists, a principle has appeared. The name "anthropic principle" was originally proposed in 1973 by Brandon Carter, a physicist at Meudon Observatory near Paris. It states that our existence as intelligent creatures constrains the universe to be what we observe, namely if just one of the fundamental constants of nature were different, life would not have developed and we would not exist. The notion of an anthropic principle was further developed by John Barrow and Frank Tipler in their 1986 book *The Anthropic Cosmological Principle*. Barrow and Tipler maintain that there are two kinds of anthropic principles. The weak anthropic principle (WAP) says that the observed values of all physical and cosmological quantities are not equally probable; they are restricted by the requirement that there must be places where carbon-based life can evolve. The strong anthropic principle (SAP) says that the universe is such that it allows life to develop within it at some stage in its history. In common usage among physicists, the anthropic principle means that the constants and the laws of physics are the way they are because otherwise human beings would not exist and we would not have a consciousness with which to perceive the universe and its physical laws. The anthropic principle paints a picture of the universe that is as human-centered as was Aristotle's pre-Copernican, Earth-centered view.

Steven Weinberg published an article in 1987 in *Physical Review Letters* in which he claimed that the cosmological constant could not be much bigger than observations allow, nor could it be much smaller, for otherwise it would prohibit the formation of galaxies and the evolution of life. His article and similar articles by Alexander Vilenkin have led to a heated debate among physicists as to how far one should rely on the anthropic principle to investigate the nature of the universe. Many physicists, including myself, consider the application of the anthropic principle in physics anathema to be avoided at all costs. Pursuing the anthropic principle as a fundamental basis for our understanding the universe is, in my opinion, a dead end, for it amounts to a tautology. You can neither test nor falsify any so-called predictions of the anthropic principle, and therefore it defeats the fundamental purpose of studying physics and understanding nature.

To return to strings: The very large or even infinite string landscape is made to appear reasonable by saying that there is an infinite number of universes, together called the "multiverse" or, as Susskind calls it, the "megaverse." String theorists say that our universe is just one of them and necessarily exists because we are here to observe it. Following the anthropic principle, if one of the constants such as the cosmological constant or the gravitational constant were significantly different from the measured values we have found in our laboratories and in astronomical observations, then our universe would not exist in the way we know it and neither would we. Since supposedly none of these infinite universes in the multiverse are causally connected, if indeed they exist, we will never be able to know what any of the others look like.[4]

This current crisis in string theory has spilled over into a veritable "blog war" on the Internet. The pros and cons of the anthropic principle, the multiverse, and the landscape are exchanged among bloggers every day. Many antistring theorists feel that if we physicists pursue this antiscience approach of the string landscape, then the advocates of intelligent design and other extreme religious and anti-Darwinian movements will grab onto this new pseudoscientific methodology to support their own spurious claims for radical changes in the way we view scientific facts.

In the end, it is left to the philosophers of science to debate the merits of the multiverse versus the "one universe" interpretation of the quantum world. Perhaps the verdict only amounts to some people preferring vanilla ice cream and others chocolate!

THE LOST GENERATION

Some of the brightest minds in physics have devoted the last twenty or thirty years to searching for a way to make string theory a predictive theory, and producing convincing evidence experimentally that the theory is sound and describes nature. After so many years, and after the waste of so much talent, it now seems improbable that string theory will ever succeed. With their mathematical training, string theorists could easily turn their attention to other routes to a unified field theory and quantum gravity.

Indeed, these bright minds constitute what can be called a "lost generation" of theoretical physicists. Much of the financial support for theoretical physics as a whole has gone into string theory over the last twenty years. Government grants and university hiring policies have steered physics toward the development of string theory. Compared to past centuries of scientific endeavor, the amount of wasted talent is staggering. This is because of the large sizes of groups devoted to the single-minded pursuit of string theory. Quantum mechanics, in comparison, was developed in the early twentieth century over a period of ten years by a mere handful of physicists such as Einstein, Schrödinger, Bohr, Heisenberg, Pauli, Born, Dirac, Jordan, and others. Today, at any given time, perhaps 1,000 to 1,500 theoretical physicists are doing research on string theory, publishing papers, and citing each other's papers, thereby skewing the whole literature citation index to make it appear that string theorists produce the most important research papers today.

One truth we have been able to count on concerning scientific pursuit over the centuries has been that only testable theories survive the intense scrutiny of experimental science. All those attempts that do not meet the tests eventually wither on the vine. Time will show how string theory fares.

CAMP TWO: LOOP QUANTUM GRAVITY (LQG) AND OTHER WAYS TO QUANTIZE GRAVITY

While the controversy about the success or lack of success of string theory has raged, some theorists, led by Abhay Ashtekar at Pennsylvania State University, Carlo Rovelli at the University of Marseille in France and Lee Smolin at the Perimeter Institute in Canada have pioneered attempts to formulate a quantum gravity theory without pursuing the grander goal of a unified theory. They emphasize that in Einstein's general relativity and in consistent modified gravity theories, spacetime geometry evolves dynamically.

One of the perceived problems of string theory as it is now formulated is that it is based on a fixed, static geometrical description of background spacetime. The strings are conceived as vibrating against the background spacetime geometry. The string enthusiasts maintain that they should work with fixed spacetimes and obtain results. Then eventually someone will figure out how to make string theory fundamentally independent of the choice of background spacetime geometry, as demanded by general relativity. The proponents of a fundamental quantum gravity theory, on the other hand, claim that string theory has already failed by not adapting itself to one of the most important features of Einstein's gravity theory: a dynamical geometry of spacetime. They consider developing a quantum gravity theory independent of a fixed spacetime background to be a primary goal.

Recall that one of the aims of quantum gravity is to reduce the gravitational waves predicted by Einstein's gravity theory to quantum parcels of energy called "gravitons." As a starting point for this pursuit, quantum gravity researchers have devised several ways to quantize spacetime. They claim that we should actually quantize spacetime itself and not just treat gravity as a field like the electromagnetic field, because as we know, in general relativity, gravity *is* the geometry of spacetime. Think of hockey players skating around in an ice rink. The players can be compared to the particles in standard field theory, while the ice is spacetime. We quantize the hockey players and obtain the very successful model of particle physics. Now we have to somehow quantize the ice itself, breaking it down into its smallest units. Thus the graviton can be pictured as one of the hockey players as well as the basic unit of the ice rink. The quantized geometry of spacetime must lead at large macroscopic scales to the classical gravity theory of Einstein's general relativity or to a physically satisfactory modified version.

One of the best known quantum gravity theories is Loop Quantum Gravity (LQG). Kenneth Wilson of Cornell University, who was awarded the Nobel Prize in 1982 for his research on critical phenomena in condensed matter physics, proposed what are now known as Wilson loops in 1970. In technical terms, these mathematical loops guarantee that the gauge symmetry invariances in quantum field theory and the symmetry of general covariance in general relativity and modified gravity theories are rigorously upheld in calculations.[5]

LQG describes spacetime geometry as a *spin network*, which is a graph with edges labeled by representations of some mathematical group and vertices labeled by intertwining operators. The spin networks and their related *spin foam* formalisms give an underlying intuitively compelling picture of LQG. These formalisms provide the possibility of obtaining a truly quantum mechanical description of the geometry of spacetime. Generically any slice of a spin foam produces a spin network.*

---

* Quantum gravity, perhaps because its modern interpretation is still evolving, demands a deep mathematical and physical understanding.

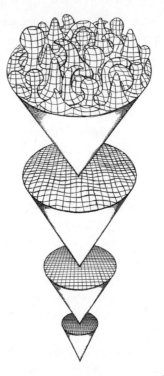

FIGURE 15: **The magnification of spacetime reveals quantum foam at the Planck length.**

The idea is that at the infinitesimal distances of the Planck length, spacetime ceases to be what it appears to us to be in everyday life. John Wheeler poetically describes this close-up of spacetime as the "foaming rush of waves on a seashore," and the mathematical jargon of spin networks is an attempt to make sense of this chaotic, fluctuating foam.

Unfortunately, the physics soon becomes highly speculative and completely divorced from any laboratory experiments that we can perform to discover the nature of the "foam." Can something emerge from this foaming speculation that can look like our spacetime when it is extrapolated to the much larger distance scales of everyday existence and laboratory dimensions?

There are several problems with the quantum gravity program. For one thing, it has been difficult to prove that the spin networks and spin foam descriptions of spacetime can, in a well-defined limit, yield a correct classical gravity theory that agrees with macroscopic observations such as the anomalous perihelion advance of Mercury. These loop quantum gravity and spacetime foam scenarios depict spacetime as a discrete set of cells with the tiny edge of a cell being equal to the Planck length. As observers, we would need a giant magnifying glass to see the tiny, submicroscopic cells. It would be like sitting in the middle of our galaxy and trying to see a fly crawling across a leaf on a tree on Earth. To get from the submicroscopic cells to the classical theory of gravity, we have to move so far away from them in distance that in effect they shrink to zero. Technically this is far from easy to achieve: One can claim to have quantized spacetime geometry and have a quantum gravity theory, but that does not prove that it is the *correct* quantum gravity theory that in macroscopic terms becomes general relativity or a modification of general relativity that agrees with the classical observational tests in the solar system, astrophysics and cosmology.

Steven Weinberg proposed a different approach to quantum gravity in the 1980s, which he called "asymptotically safe quantum gravity." This theory is an attempt to quantize the gravitational field, or gravitational waves, rather than spacetime per se. The idea of asymptotic safety or, as others have called it, "asymptotic freedom," originated in particle physics in the quantum field theory interpretation of the strong forces between quarks that bind atomic nuclei. As the energy of the quarks increases and the distance between them decreases, the strong force between the quarks decreases and at very high energies it eventually vanishes, leaving the quarks as free particles. The huge energies needed to make the quarks free cannot be achieved in Earth-based accelerators, but may have been present in the very early universe.

"Asymptotic" means approaching ever closer. In this case, the quarks are approaching a state of vanishing interaction with each other as the energy increases, and so they become "asymptotically free." This state is reached when the strong force coupling strength becomes zero, which happens at extremely high energies. We do not observe free quarks in accelerators because as the distance between them increases or conversely, as the energy decreases toward the low energies in accelerators, then the strong quark force begins to confine the quarks inside the proton or neutron, preventing them from becoming free and visible in the laboratory. This anti-intuitive situation is called "quark confinement."[6] The discovery of asymptotic freedom in the quantum field theory of strong quark interactions was made by David Gross, Frank Wilczek, and David Politzer, who were awarded the Nobel Prize for physics in 2004.

Weinberg conjectured that asymptotic safety might also be applicable to quantum gravity. As in the case of the coupling strength of the strong force becoming weaker as the energy increases, he reasoned that the same could happen with Newton's gravitational constant G. The variation of G can have significant consequences for astrophysics and cosmology. G as a function of energy would decrease to zero asymptotically as the energy between two gravitating bodies increases to infinity, or equivalently, it would decrease to zero as the distance between two gravitating bodies approaches zero. On the other hand, as the energy of gravitating bodies goes to zero, or as the distance between them increases, then the strength of gravity *increases.*

As we shall see in Chapter 10, this increase of gravity with distance is a major characteristic of MOG and the reason that MOG fits astrophysical and cosmological data. That is, the situation at large galactic distances is very similar to the situation at microscopic levels in that the confining gravitational force can produce the observed flat rotation curves of spiral galaxies and explain the stability of X-ray clusters of galaxies without dark matter. Weinberg also conjectured that in an asymptotically safe quantum gravity theory, all the problems of infinities in quantum gravity calculations would be removed, leading to a successful theory of quantum gravity. This approach has been pursued vigorously by some physicists over the past two decades. A leader is Martin Reuter and his collaborators at the University of Mainz, Germany.

Another approach to quantizing spacetime geometry has been to apply quantum mechanics to the coordinates of spacetime by postulating that they do not commute. That is, when you multiply four coordinates of a point in spacetime by another four coordinates in spacetime, you get a certain answer. When you multiply the two sets of coordinates in reverse order, you get a different answer, so that when you take the difference of the products, you get a nonzero answer. This is what noncommutativity in spacetime means.[7]

In quantum mechanics, the order in which two physical quantities are multiplied together is important. For example, when the coordinates of the position of a particle in space are multiplied by the momentum of the particle and then the reverse order of this product is subtracted, the result does not vanish as it would for the classical values of the coordinates of the particle position and momentum. This is one of the most fundamental features of quantum mechanics and underlies Heisenberg's famous uncertainty principle that the momentum and position of an electron, say, cannot both be determined with infinite precision by observation. In noncommutative quantum gravity, this lack of commutativity of the coordinates naturally brings in a finite length, which it is hoped will play a role in making the quantization of spacetime lead to a finite quantum gravity theory without infinities in calculations.

WHAT ABOUT THE DATA?

At various times I have been involved with the quantum gravity approach—through a so-called finite, nonlocal quantum gravity theory and also noncommutative quantum gravity[8]—and I believe there are serious problems. For one thing, to date not one of the experimental tests that has been proposed to verify (or falsify) any of the quantum gravity theories, including my nonlocal quantum gravity theory, has been conclusive. What is more disconcerting is that none of these proposed tests can distinguish one quantum gravity theory from another. Of course, these problems plague string theory as well.

Secondly, it is highly doubtful that the graviton itself will ever be detected. This constitutes a very big obstacle to creating a theory that combines gravity and quantum mechanics. The difficulty is that gravity operates over such large masses and distances, whereas quantum mechanics takes place in the very peculiar realm of the almost inconceivably small. At atomic and nuclear distance scales, the force of gravity is about 10^{40} times smaller than the strength of the electromagnetic force. This means that the gravitational force between electrons and the atomic nucleus is about 10,000 billion billion billion billion times weaker than the electrical attraction. This is why in the standard particle physics model, which agrees with all known experiments to date, gravity is neglected; it simply doesn't matter quantitatively in calculations at atomic and nuclear scales. In order to see the effects of quantum gravity—or gravity at the submicroscopic level—we would have to go to enormous energies, namely the Planck energy of about 10^{28} electron volts. To reach this enormous amount of energy, we would have to build such a huge high-energy accelerator that its circumference would be the size of the Milky Way galaxy. Only at such enormous energies does the force of gravity become comparable to the electromagnetic force, which would make the graviton detectable by experiment.

Another way of looking at the weakness of gravity is to think about the gravitational effect on hydrogen atoms. Here the additional attraction of gravity between the electron and the proton in the hydrogen atom—over and above the strong electromagnetic attraction—would create a small change in the energy of the hydrogen atom. In quantum mechanics, the energy of the hydrogen atom wave function corresponds to a frequency of about 10^{16} cycles per second. In order to observe any gravitational effect on the hydrogen atom's energy, we would have to wait a very long time indeed to see any effect: 100 times the age of the universe! Clearly, gravity has a very small effect in atomic physics.

What about testing quantum gravity by finding the graviton in more conventional experiments on Earth? Recall Einstein's 1905 photoelectric effect, which proved the existence of the photon. It was eventually shown experimentally that it is the frequency of light in the form of X-rays hitting the electrons in a metal and ejecting them that proved that light consisted of the "corpuscles" of energy we now call photons. One can see the experimental evidence of the scattering of the photons and electrons in a cross-section, a measure of the scattering area.*

It is almost inconceivable that a similar scattering experiment could be done to prove the existence of the graviton. Because the force of gravity is so weak, the cross-section of gravitons colliding with electrons would be incredibly tiny—about 10^{-66} centimeters squared. Such tiny areas are completely beyond our ken. As for the frequency of graviton-electron collisions, a graviton would manage to hit an electron perhaps only a few times in the lifespan of the universe—making a detection of the graviton virtually impossible.

Why do we try to construct a quantum gravity theory that is based on a quantum of energy called the graviton that it seems we can never detect? Does this mean that any quantum gravity theory is untestable? The only other way we can conceive of testing quantum gravity is to seek some experimental signature from the very early universe when energies may reach the Planck energy. This has led physicists to speculate upon effects in cosmological observations of the early universe when the densities of matter and energy, and temperatures, were very high. But so far no such definitive signatures have been discovered.

Some theorists simply claim that since gravity is observed and quantum mechanical effects are also observed, that qualifies as enough experimental evidence that a quantum gravity theory is necessary. Such theorists, who have devoted years of their lives to the quest for a theory of quantum gravity, claim that on philosophical grounds alone, you have to quantize the gravitational field and make it consistent with the successful theory of quantum mechanics and quantum field theory—and someday, somehow, evidence for the truth of the theory will pop up. Either that, they claim, or such proof simply does not matter.

---

* Three Nobel Prizes resulted from this work. After the prize was awarded to Einstein in 1921, it was bestowed on Thomas Millikan in 1923 for demonstrating the photoelectric effect in the laboratory, and on Arthur Compton in 1927 for his work on the change in wavelength of X-rays when they collide with electrons or protons. This latter phenomenon, the Compton effect, confirmed the dual nature of electromagnetic wave and particle.

Chapter 9

OTHER ALTERNATIVE GRAVITY THEORIES

I n the previous chapter we considered how to explore gravity at the very small, microscopic scales, when gravity is described by quantum language. In particular, we expect that quantum gravity becomes important when energies approach the Planck energy of 10^{28} electron volts or distances the size of the Planck length of 10^{-33} centimeters. Now we shall turn to alternative gravity theories at the macroscopic, classical scales, which apply to the solar system and far beyond.

I published my first modified gravity theory (Nonsymmetric Gravitation Theory, or NGT), in 1979. The notion of modifying Einstein's theory as a classical theory of gravity has recently become popular because of the disturbing conundrum of the invisible dark matter and dark energy. To physicists such as myself, the huge amount of invisible dark matter needed to make Einstein's theory fit the astrophysical data is reason enough for exploring modified gravity theories. Also, the claim that the expansion of the universe is accelerating has created a whole new impetus for discovering a classical, as opposed to quantum, modification of Einstein's gravity theory. If in the future dark matter is not detected, and if the dark energy associated with the accelerating universe also remains unexplained, then we would have to accept a modified gravity theory at the macroscopic and classical levels. This new theory could also become the basis of a quantum theory of gravity.

In this chapter I will describe attempts by other physicists to modify Einstein's gravity theory. The following chapter will focus on MOG.

THE CHALLENGES

Let us be clear at this point: It is exceedingly difficult to succeed in any attempt to modify Einstein's gravity theory. Modified gravity theories constitute a graveyard. When walking through this graveyard, we view the many headstones on which are inscribed the names of authors of failed theories.

A modified gravity theory has to be at least as self-consistent mathematically and physically as Einstein's gravity theory. Otherwise it is doomed to failure from the beginning. One of the serious challenges facing a modified gravity theory is the issue of physical instability of one form or another that can arise out of the basic equations of the theory. Such instabilities can lead to the embarrassing prediction by a theory that the sun would only shine for two weeks, or that the universe would only exist for 10^{-26} years, which is one-third of a trillionth of a microsecond! Instabilities could take the form of negative energy solutions of the field equations, and other pathologies. Negative energies are generally considered to be anathema because they have never been observed. Almost all modified gravity theories can be excluded for reasons of instabilities.

Instabilities in modified gravity theories are like deadly viruses that a patient contracts. The doctor pronounces there is no cure. Eventually the effects of the virus begin to diminish, and the patient goes into remission, to survive a while longer. However, the unphysical modes of the theory get triggered at some point and finally kill the patient. I have often wondered whether Einstein realized that he was constructing such a robust theory of gravity in the form of his general relativity.

A successful modified gravity theory must agree with the following observational data:

1. All the accurate observational data associated with the planets of the solar system, including Earth-based measurements of Einstein's equivalence principle (the Eötvös experiments) and the Cassini spacecraft measurement of the time delay of radio signals*
2. Data from the binary pulsar PSR 1913+16
3. The rotation curves of galaxies
4. The observed mass profiles of X-ray galaxy clusters
5. Weak and strong gravitational lensing of galaxies and clusters
6. The lensing data of merging clusters

---

* The Cassini is a NASA mission that orbits Saturn; it has obtained data that determine a maximum value for the possible variations of G with time.

7. The cosmic microwave background data, in particular the acoustical waves in the CMB power spectrum
8. The formation and growth of galaxies beginning with the early universe, as revealed by the matter power spectrum data based on large-scale galaxy surveys
9. The observed apparent acceleration of the expansion of the universe as determined by supernovae data

Any modified gravity theory that fails to describe all of these observational data without using dark matter should be abandoned, and its authors should begin designing their headstones.

JORDAN-BRANS-DICKE GRAVITY

The grandaddy of alternative gravity theories is the Jordan-Brans-Dicke theory, developed by Pascual Jordan in the 1950s and later by Carl Brans and Robert Dicke in the early 1960s. One of the motivations of these authors was to incorporate Mach's principle into gravitation. Recall that this principle states that the inertial mass of a body is due to all the rest of the matter in the universe. A fundamental explanation of the inertia of a body and its orgins would be an attractive feature of a gravity theory. As we recall from Chapter 2, Einstein eventually gave up attempting to incorporate Mach's principle into general relativity. Whether Mach's principle is successfully incorporated in Jordan-Brans-Dicke theory is still a matter of controversy.

In any case, in order to incorporate Mach's principle into their modified gravity theory, the authors had to introduce a varying gravitational constant. Paul Dirac's idea of varying the gravitational constant G with time to explain the weakness of the gravitational force, published in *Nature* in 1937, inspired the German physicist Pascual Jordan to publish a modified gravity theory in *Nature* in 1949. He wrote a more detailed presentation of the theory in his book *Schwerkraft und Weltall*, published in Germany in 1955. In Jordan's gravity theory, the variation of Newton's gravitational constant with space and time was described by a simple scalar field. However, others criticized the theory, including Markus Fierz and Hermann Bondi of steady-statesmen fame, because the energy-momentum tensor of matter and therefore energy were not conserved. Taking such criticism to heart, Jordan published an improved version of the theory in 1959, but the theory still had some problems with its treatment of matter.

Carl Brans and Robert Dicke of Princeton University published a more complete theory of gravity incorporating a varying gravitational constant in the *Physical Review* in 1961, followed the next year by a second paper by Dicke. In this modified gravity theory, called in the physics literature the "Jordan-Brans-Dicke scalar-tensor-gravity theory," the gravitational constant G is replaced by the inverse of a scalar field, and the strength of the coupling of the scalar field to matter is measured by a constant identified by the Greek letter ω (omega). Due to the significant improvement in the accuracy of gravitational experiments over the past half-century, this constant now has to be greater than 10,000 so that its inverse, which appears in the gravitational field equations, is extremely small. Variations of scalar-tensor-gravity have been proposed more recently, such as providing the scalar field with a mass, making its interactions with matter only significant at short distances.

The Jordan-Brans-Dicke theory is no longer considered a contender in the alternative gravity theory arena as far as the solar system observations are concerned, because the coupling strength that measures the size of the extra scalar field is required to be so small that it cannot be detected in the solar system. However, Jordan-Brans-Dicke scalar-tensor gravity theory has been a popular vehicle used by cosmologists to describe the early universe.[1]

MILGROM'S MOND

In 1983, the Israeli physicist Mordehai Milgrom published a phenomenological model in which he modified Newtonian gravitation by saying that when the gravitational acceleration acting on a body falls below a certain critical threshold, some ten orders of magnitude smaller than the gravitational acceleration we experience on Earth (9.8 meters per second per second), then the Newtonian law of gravitation is modified. The nonrelativistic formula of Milgrom's Modified Newtonian Dynamics (MOND) says that when the acceleration of a body is greater than the critical acceleration $a_0 = 1.2 \times 10^{-8}$ centimeters per second squared, then Newton's law of gravitation applies, and when it is less than that critical acceleration, then the MOND formula kicks in.

Milgrom's motivation for this work was to remove the need for exotic dark matter by modifying Newton's law of gravity. He fitted MOND to galaxies and found, surprisingly, that it fitted remarkably well the data for the rotational velocities associated with the circular orbits of stars and gas in the galaxies. No dark matter was needed to help fit the data. Milgrom also made some other predictions in subsequent papers, which more or less turned out to be true when compared to observational data.

However, there are theoretical problems with MOND. First, the model is not a fully relativistic theory of gravity. That is, it is not consistent with Einstein's general relativity. Secondly, in Milgrom's original model the momentum of a particle, that is, its mass times its velocity, is not a conserved quantity. This is equivalent to giving up the conservation of momentum and energy, which are fundamental, sacred principles of physics. However, in the early 1980s, the Israeli physicist Jacob Bekenstein and the American physicist Robert Sanders succeeded in producing a nonrelativistic version of MOND that avoided violating the conservation of momentum. A third problem with MOND is that there is an uncertainty in the magnitude of the acceleration of a body when it is moving in a field of acceleration caused by other bodies. For example, a particle moves in the acceleration field of a planet, but the planet itself is moving in the acceleration field of the whole solar system. So which acceleration do we use in the MOND formula? Thus, when is the acceleration of a body greater or smaller than the MOND critical acceleration?

Despite these problems, the good fits to galaxy rotation curves have spurred much interest in Milgrom's model, and MOND has received a lot of attention from astronomers. Because it is a phenomenological model, not a relativistic gravity theory, the mathematics are easy to understand. Although recently it has been shown not to work at all with galaxy clusters without some form of dark matter, a veritable industry of postdocs in Europe, UK, and North America is still busy attempting to fit MOND to astronomical data.

Stacy McGaugh of the University of Maryland is one of the astronomers who initially was most enthusiastic about MOND. When I first met Stacy five years ago when he visited the Perimeter Institute, he told me that as a graduate student vigorously pursuing the dark matter solution to the problem of the motion of stars in galaxies, he had firmly believed in the existence of dark matter. "But by the mid-nineties, the dark matter picture was becoming loaded down with fine-tuning problems," Stacy said. By chance, Stacy had heard Milgrom speak on MOND, and although he was dubious, he read Milgrom's original papers. Surprised, he found that Milgrom had predicted what Stacy had later observed. "It was an epiphany!" Stacy said excitedly. "MOND naturally explained what was confusing in dark matter. I realized that MOND could fit the galaxy data in a way that dark matter could never do, without too many embarrassing complications and adjustable parameters, and this altered the course of my research. I felt that MOND had a lot going for it, and this was the way to resolve the galaxy problem without undetected dark matter." He did add that he knew that MOND had some problems, and he hoped they would be resolved in the future.

MOND ran into serious difficulties with the publication of the so-called bullet cluster data in 2006 because it could not fit the data without dark matter (see Chapter 10). Similarly, and earlier on, Robert Sanders, a senior member of the Kapteyn Observatory in Groningen, Holland, and one of MOND's champions, showed that it was not possible to use Milgrom's MOND acceleration formula to fit the mass profiles of normal, noncolliding X-ray clusters without some form of dark matter. Sanders suggested that perhaps one could ameliorate the situation with massive neutrinos. However, a stable electron neutrino with a mass of two electron volts was needed to fit the data, which is marginally in agreement with bounds obtained from current neutrino experiments and cosmological observations. Indeed, my graduate student Joel Brownstein and I attempted to fit Milgrom's MOND formula to the X-ray clusters, and discovered that it did not fit the data unless we assumed a significant amount of dark, nonbaryonic matter. Recent work by Ryuichi Takahashi and Takeshi Chiba at Kyoto University has confirmed that Milgrom's MOND cannot fit the X-ray cluster mass profiles or the gravitational lensing data without dark matter. They showed that even changing the value of Milgrom's critical acceleration constant or including neutrinos with mass could not save the situation.

If MOND supporters cannot find a means of resolving these problems without dark matter, they should turn to designing MOND's headstone.

BEKENSTEIN'S RELATIVISTIC VERSION OF MOND

For more than twenty years since the publication of Milgrom's first paper on MOND, Jacob Bekenstein and Robert Sanders have worked on making a consistent relativistic theory of gravitation out of Milgrom's MOND phenomenological formula. They have been trying to overcome some of the model's fundamental problems. Part of the task of constructing a gravity theory that can solve the dark matter problem—that is, do away with the need for exotic dark matter entirely—is being able to explain the deflection of light rays and the lensing phenomenon for very distant quasars. As we have seen, lensing occurs when light from a quasar or bright galaxy travels near a strong gravitational field like a galaxy or a cluster of galaxies closer to us, and the path of its light is bent, as the large object acts like a lens. Only a relativistic theory of gravitation, such as Einstein's gravity theory, can fully explain the lensing phenomenon. Another challenge in creating a new gravity theory is that only a fully relativistic gravity theory can explain the large-scale structure of the universe—how the galaxies and clusters of galaxies formed—and be consistent with the growing amount of accurate cosmological data accumulating every year.

Finally, in 2004, Bekenstein published a paper in *Physical Review D* claiming to have succeeded in making a consistent relativistic theory out of MOND. This theory, called Tensor-Vector-Scalar (TeVeS) gravity, contains two metrics in spacetime, and so constitutes a so-called bimetric theory of gravity of the kind described in Chapter 6 in relation to VSL. In addition, the theory contains two scalar fields and a vector field.* One of the scalar fields occurs in an arbitrary function chosen to yield Milgrom's MOND formula. This allowed Bekenstein to reduce his theory to the nonrelativistic MOND acceleration law.

A theoretical problem with Bekenstein's modified gravity theory based on MOND is that the vector field that is part of the theory's mathematical structure has some pathological physical consequences. The vector field was needed in the theory to produce enough bending of light beyond that predicted by Einstein's gravity theory to explain the lensing phenomenon of galaxies and clusters without dark matter. Technically speaking, one of the consequences is that the vector has to be a timelike vector, and Bekenstein achieves this in his action principle by introducing what is called a "Lagrange multiplier field." A direct consequence of this is that the theory violates local Lorentz invariance and predicts preferred frames of reference. Therefore TeVeS is not strictly speaking a fully relativistic theory because of the manner in which the vector field violates Lorentz invariance. The theory reverts back to the ether theories in vogue before Einstein published special relativity in 1905.

Another consequence of the vector field is that the energy of a physical system is not bounded from below. This means that the system does not have a state of lowest energy, which is unacceptable for a physical system. And so the problems of instabilities and negative energies, which have never been seen on Earth, rear their ugly heads again. Finally, and more recently, Michael Seifert, a graduate student working with Robert Wald at the University of Chicago, has published a paper in *Physical Review D* proving that the TeVeS solution for the solar system leads to a sun that would shine for only two weeks!

Yet Bekenstein's theory has attracted attention. Constantinos Skordis at the Perimeter Institute, together with Pedro Ferreira and collaborators at Oxford University have attempted to describe the cosmic microwave background data in terms of cosmological solutions of Bekenstein's theory. A

---

* A scalar field, such as the Coulomb field potential caused by electric charge, has no direction in three-dimensional space. But a vector field does, such as the electric and magnetic force fields united in Maxwell's equations of electromagnetism.

potential problem occurs when attempting to obtain from the theory a satisfactory fit to the acoustical power spectrum data obtained by the WMAP team and in balloon experiments that I discussed in Chapter 7. Dark matter, in the form of neutrinos with a mass of two electron volts, must be included. Finally, Bekenstein's theory has so far not provided any explanation for the dark energy and the accelerating expansion of the universe.

Despite the problems inherent in TeVeS, Bekenstein and Sanders have shown tenacity and courage in trying to develop a modified gravity theory over almost a quarter of a century, particularly in the face of continual skepticism and hostility from the majority of the astrophysics community.

MANNHEIM'S CONFORMAL GRAVITY

Another way to pursue a modified gravity theory is to stay within a purely geometrical structure without adding additional fields. Philip Mannheim of the University of Connecticut, an imaginative and skilled physicist, published a modification of Einstein gravity using what is called "conformal gravity" in the early 1990s. This theory is fully relativistic. Mannheim's original motivation in constructing his theory of gravitation was to make quantum gravity renormalizable (finite). Later he discovered that it could get rid of dark matter as well.

A conformal theory is one that does not possess a fundamental length generic to the theory. Einstein's gravity theory and my MOG are not conformal theories of gravity because the matter and energy of spacetime are not zero everywhere. In contrast, in Mannheim's conformal theory, all particles are massless. An example of a conformal theory is Maxwell's electromagnetic theory describing the electromagnetic fields, because the photon has no mass. In Mannheim's theory, not only is the graviton massless, but *all particles* are massless. Therefore to complete his program and to make it match reality, Mannheim has to break the conformal invariance of his theory.[2]

FIVE-DIMENSIONAL MODIFICATIONS OF GRAVITY

Georgi Dvali from New York University and his collaborators Gregory Gabadadze and Massimo Porrati have proposed modifying Einstein's gravity theory in order to explain the accelerating universe. These authors employ a five-dimensional model: Our three-dimensional space would reside on a brane in a five-dimensional space called a "bulk." Gravity acts in the bulk, so at large cosmological distances, Einstein's theory is modified to account for the acceleration of the expansion of the universe. Speculative models—and there are several—such as the one proposed by Dvali and his collaborators also have consequences at the microscopic scale. Dvali's model suggests that Newton's inverse square law breaks down for distances less than a thousandth part of a millimeter: There must be radical modifications to Newton at that distance scale.

One of the worst determined constants of nature is Newton's gravitational constant. Much effort has been devoted over many years to improving its experimental value by testing Newton's inverse square law of gravity, which we recall states that gravity decreases as the square of the distance between two objects. Newton's gravitational constant is difficult to measure because of the weakness of the gravitational force. It is known from spacecraft measurements orbiting Mars and Jupiter that the inverse square law is well determined for the planets. But what about microscopic distances?

Eric Adelberger and his collaborators at the University of Washington in Seattle have been performing ingenious experiments to test the inverse square law at small distances in an Earth-based laboratory, and relating those tests to dark energy. Their recent experimental results have led to a negative conclusion: There is no violation of the inverse square law down to a distance of a thousandth of a millimeter. Newton's law of gravity holds.

In addition to this potential falsification of Dvali's model by experiment, multidimensional models such as this are subject to possibly severe instability problems, which has been the source of much controversy over the past several years. Neither does the model solve the coincidence problem— namely, why is the acceleration happening now?—without significant fine-tuning of its parameters.

WEIGHING THE ALTERNATIVES

The number of classical modifications of Einstein's gravity theory increases year by year. More and more physicists are becoming motivated by the conundrum of dark matter and dark energy to see whether modified gravity alone could account for the data. Indeed, when I attend international conferences, often a colleague will approach me and say enthusiastically, "Hello, John! Have you seen my latest published modification of Einstein gravity?" I can't help but think of my early days in the late 1970s and early 1980s when I struggled with referees whenever I submitted papers on my own modified gravity work to journals.

Today, with several modifications of general relativity being developed— and I have described some of the important ones aside from MOG in this chapter—there has to be some way of separating the wheat from the chaff.

As I have stressed, one important criterion is that theories must not have pathological instabilities in the form of negative energies. In fairness to Einstein, we need to make sure that the modified theory we develop is at least as physically consistent as general relativity before we start trying to use it to fit astrophysical and cosmological data. Also, it is now becoming clear that the many recent observations of lensing of galaxies and clusters of galaxies are producing a better understanding of the distribution of matter in the universe. A modified gravity theory has to be relativistic in order to account for the considerable amount of lensing data being published, for the lensing phenomenon is a purely relativistic effect due to the curvature of spacetime. These data will become crucial in determining which alternative gravity theory can succeed in getting rid of dark matter.

The modified gravity theories and models that I have discussed in this chapter have problems—whether of physical instabilities or fitting solar system data. With strings, quantum gravity and these other alternative gravity theories as background and context, let us now turn our attention to MOG.

Chapter 10

MODIFIED GRAVITY (MOG)

There are three main reasons for attempting to modify Einstein's general relativity. First, many theorists are convinced that we have to find a quantum gravity theory that combines general relativity with quantum mechanics, and many suspect that a successful quantum gravity theory would only come about through a fundamental modification of Einstein's classical gravity theory. Second, the amazingly successful standard model of particle physics does not contain the gravitational force, and physicists hope that a larger theory containing the standard model of particle physics as well as gravity will emerge. Third, we need to understand this serious problem of the missing mass and energy in the universe. It is this third motivation that I will focus on in this chapter. Just as Einstein in his day was constructing an alternative gravity theory—an alternative to Newton's, which had prevailed for more than 200 years—so I have been searching for a larger theory, a modification of general relativity that would fit the data without the need to posit dark matter, and would contain Einstein's theory just as Einstein contains Newton.

Unlike the alternative gravity theories discussed in the previous chapter, my mature Modified Gravity theory, MOG, contains no physical instabilities. It is as robust a gravity theory as general relativity, and fits all the current astrophysical and cosmological data without dark matter.

BUILDING THE NONSYMMETRIC GRAVITATION THEORY (NGT)

There are two possible ways to modify Einstein's gravity theory in four-dimensional spacetime. One is to somehow keep within the framework of

Riemannian geometry and Einstein's gravity theory, and attempt to modify how matter interacts with spacetime geometry—as in Mannheim's conformal gravity theory. The other is to introduce new fields—as in Bekenstein's TeVeS—or some new kind of geometrical structure that extends Riemannian geometry. These latter kinds of modifications are based on what is called "non-Riemannian geometry." My Nonsymmetric Gravitational Theory (NGT) is a theory of this type, and it contains Einstein's gravity theory.

In 1979 I published a paper in *Physical Review D* entitled, "A new theory of gravitation," which was based on the mathematics of Einstein's nonsymmetric unified field theory. Toward the end of his life, Einstein had been attempting to construct a theory that unified gravity and Maxwell's equations of electromagnetism. However, several physicists, including myself, showed that Einstein's unified theory did not in fact correctly describe Maxwell's theory. But in 1979 I reinterpreted Einstein's nonsymmetric unified theory as a generalization of his purely gravitational theory. Recall that the metric tensor is one of the basic ingredients of Einstein's theory; it determines the distance between two points in spacetime, and thereby determines the warping of spacetime. In the nonsymmetric extension of Einstein's gravity theory, the symmetric metric tensor is replaced by a tensor that is the sum of symmetric and antisymmetric tensors. The antisymmetric part of the metric tensor has the effect of twisting the geometry of spacetime. NGT's spacetime, in addition to warping spacetime like Einstein's gravity theory, has a torsion that twists the warping of spacetime. In collaboration with my graduate students at the University of Toronto, I developed this theory extensively over a period of twenty years. During many of those years, there were no observational data that could test my generalization of Einstein's gravity theory or that could distinguish my theory from general relativity.

Einstein considered his nonsymmetric theory, which attempts to unify gravity with electromagnetism, to be the most natural generalization of his gravity theory from a mathematical point of view. He adopted this a priori reasoning in his later life. He believed that if he discovered the most natural generalization of his gravitation theory, then eventually the theory would be vindicated by experimental data. Many physicists at that time believed that Einstein was taking the wrong path in his physics research because his new theories could not be verified experimentally. This is analogous to the methods of string theorists today, who speculate on deriving a unified field theory without any observational verification.

I, too, became trapped within this mode of thinking, and spent several years searching for the ultimately perfect generalization of Einstein's gravity theory without any experimental motivation or verification to back it up.

Moreover, I hoped that the generalization of Einstein's theory would lead to a consistent quantum gravity theory, a better understanding of the beginning of the universe, and an elimination of the problems of infinities and singularities in Einstein's theory. This pursuit of a mathematically consistent vision at least led to the education in physics of many talented PhD students, and the publication of many papers coauthored with them in peer-reviewed journals. Yet always we were waiting and hoping for data to come in that would vindicate all the research we had been doing. The larger community of relativity physicists, meanwhile, ignored or derided NGT because, in the absence of new data, there just seemed to be no compelling reason for a new gravity theory. Thirty years ago, before the new cosmological and astrophysical data appeared, dark matter was not considered a pressing issue.

But for a brief time in the early 1980s, it looked like the data were coming in to verify NGT, and my hopes soared. In 1982, Henry Hill, a physicist and astronomer at the University of Arizona in Tucson, published a paper claiming that the sun was not exactly spherical in shape, but was oblate—and further, that this oblateness would cause a discrepancy with Einstein's equations predicting the precession of the perihelion advance of Mercury. Hill's paper was based on his observations of the sun using the solar telescope on Mount Lemon outside Tucson.

This new development caused much excitement in the media because it seemed that Einstein might be wrong. And it led to a furor among physicists. Hill's observations actually supported previous observations made by Robert Dicke and his collaborators at Princeton University in the 1960s, which also showed that the oblateness of the sun would cause a discrepancy with Einstein's theory, although Hill's calculations of the discrepancy were larger than Dicke's. I calculated the possible correction to the perihelion precession of Mercury from my NGT equations and discovered that I could obtain agreement with Hill's observations. Thus it might be possible to verify my new gravity theory.

In 1984, at the height of the controversy, the New York Academy of Sciences gathered together some of the world's experts on solar astronomy and relativity, including Henry Hill and myself. At this meeting, I presented the results of my calculations based on NGT. Most of the physicists at the meeting voiced much skepticism of Hill's claim that his data invalidated Einstein's theory. Articles were published in *The New York Times* and other newspapers worldwide, presenting the pros and cons of this controversy.

At that time, some solar astronomers were pursuing a different way of investigating the possible oblateness of the sun. They were using the methods of helioseismology, in which the acoustical properties of the interior of the sun were studied to determine its shape. These results purportedly showed that Hill's claims about the oblate shape of the sun were incorrect. At ensuing international meetings, further observational evidence based on helioseismology was presented, and the proponents of these observations attempted to discredit Henry Hill, claiming that he had committed errors in his analysis of the data. A consensus formed among observational astronomers that there was no correction needed to Einstein's prediction of the perihelion advance of Mercury, and this consensus holds today. The excitement that my new gravity theory could be differentiated from Einstein's died down.

For many years, I was faced with the situation described by the world's most celebrated detective, as Sherlock Holmes complained to Dr. Watson: "I have no data yet. It is a capital mistake to theorize before one has data. Insensibly, one begins to twist facts to suit theories instead of theories to suit facts."

Once the controversy had died down, was I left with an incorrect shape of the sun in my theory? No. If the sun were oblate, then there would be an extra correction to general relativity that could be accounted for in NGT. NGT contains an extra repulsive force, which is employed in the calculation of the perihelion advance of Mercury. A free, adjustable constant in NGT determines the magnitude of this repulsive contribution to the perihelion advancement formula. If the sun were perfectly spherical, then this new parameter would be too small to be observed in the solar system.

After many years of studying NGT as a gravitational theory and not as a unified theory of gravity and electromagnetism as proposed by Einstein, I feel that the present status of the theory is inconclusive. One of its problems is the possibly pathological solutions of the field equations. Instabilities can occur in the field equations. For example, when the antisymmetric fields causing the torsion of spacetime become very small, then other geometric quantities in the theory, such as those affecting the orbits of the planets, may become unphysically huge, disagreeing with the observational data.

My graduate student Michael Clayton at the University of Toronto carried out an extensive study of the possible pathologies in NGT in the late 1990s, which was the basis of his PhD thesis and several published papers. More recently, Tomislav Prokopec and his student Tomas Janssen at the University of Utrecht in the Netherlands have further studied the difficult problem of discovering possible instabilities in NGT. The application of NGT to weak gravitational fields of galaxies and clusters of galaxies in cosmology has shown that the effects of the antisymmetric part of the metric tensor and its torsion are large compared to the effects due purely to Einstein gravity. This means that the perturbative techniques applied by Clayton, Prokopec, and Janssen, in which they assume that the antisymmetric part of the metric can be made vanishingly small, may not be correct. This may invalidate the instability arguments against NGT.

NGT has a rich structure and some attractive mathematical features. However, my astronomical colleagues complained to me that the mathematical structure of NGT was quite complicated. This in itself need not be a deterrent when constructing a new theory. When Einstein published his general relativity, most of the physics community found it difficult to understand and indeed the mathematical structure of the theory was far more complicated than Newtonian gravity. However I thought it was possible to discover a simpler alternative to my nonsymmetric theory that would remove the need for dark matter and explain or remove dark energy.

BUILDING MODIFIED GRAVITY (MOG)

In 2003, I set myself the task of attempting to discover such a simple alternative to NGT.[1] My new theory, Metric-Skew-Tensor Gravity or MSTG, was different from NGT in that the spacetime geometry I used was the purely symmetric spacetime of Einstein gravity. The antisymmetric part, the "skew," of NGT was no longer part of the geometry, but now occurred as an additional field in the equations of the theory. In other words, the skewness came not from spacetime itself but from a new field in the theory that in fact constituted a new fifth force.

MSTG predicted a modified Newtonian acceleration law that could fit the large amount of anomalous rotational velocity curve data from galaxies obtained by astronomers since the late 1970s, which showed stars rotating at the edges of galaxies at twice the rate predicted by Newton and Einstein. Like other originators of theories before me, I knew about these data at the time I was working out the theory. My aim was to try to explain the data without the conventional reliance on exotic dark matter. With some rudimentary fitting to the data, I was able to succeed. This fitting of the data was in fact based on previously published work with my post-doc colleague Igor Sokolov in 1995.

In order to fit the solar system data, such as planetary motions and the anomalous perihelion advance of Mercury, I postulated that Newton's gravitational constant G varied in space and time.

When I discussed my new MSTG theory with astronomy colleagues, they told me that the mathematics of this theory were *still* complicated, although not as complicated as the mathematics of NGT. In 2004 it occurred to me that I could develop an even simpler modification of Einstein gravity that could predict all the weak gravitational field phenomena associated with galaxies, clusters of galaxies, the solar system and cosmology. That is, it could explain the surprising speed of stars in galaxies, could account for the "extra gravity" that must exist in galaxy clusters, and could explain the new wealth of cosmological data. This third theory was based on a symmetric Einstein metric, a vector field that I called the "phion field," and three scalar fields. The phion particle is the carrier of the theory's fifth force, while the scalar fields describe the variation of the gravitational constant, the coupling of the phion field to matter, and the effective mass of the phion field.[2] I called this gravity theory Scalar-Tensor-Vector Gravity (STVG) and was able to prove that it was stable, had no negative energy modes, and was free of any pathological singularities.

Finally, I felt that I was getting somewhere. The variation of the gravitational constant, described by a scalar field in my theory, had much affinity with the Jordan-Brans-Dicke scalar-tensor-gravity theory.

The reader may be wondering why I was now in possession of three theories that constituted a MOG (modified gravity theory) containing Einstein's general relativity. The three theories were indeed found to be progressively simpler in their mathematical structure. But they can only be distinguished from each other by their predictions for strong gravitational fields, while their predictions for weak fields produce the same physical predictions for the solar system, astrophysics and cosmology. Strong gravitational fields are not easily come by. They could have existed in cosmology at t = 0 as well as in the gravitational collapse of stars. These phenomena would be extremely challenging to use as verifications for one or more of my theories, and no other, more accessible examples of such strong gravitational forces are available at present. What was I to do with my three apparently indistinguishable and unverified theories?

At this perplexing stage, in the summer of 2004, I was approached by Joel Brownstein, who had completed a master's degree in physics under my supervision at the University of Toronto in 1991. At that time, we had collaborated on a paper published in the *Physical Review* describing NGT predictions for the Gravity Probe B gyroscope experiment under the leadership of Francis Everett of Stanford University. Joel asked me whether I could supervise him toward the PhD in physics at the University of Waterloo, Ontario, just up the road from my office at the Perimeter Institute. During the previous decade, Joel had been working as a software engineer in industry and had acquired considerable knowledge in computing.

In Waterloo, Brownstein and I set about developing a means to fit MSTG and STVG to a large amount of rotational velocity data for galaxies without

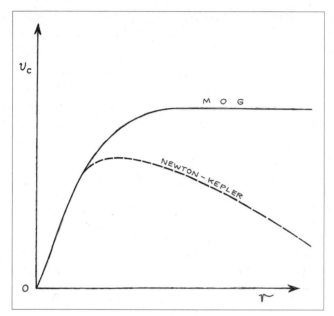

FIGURE 16: **Galaxy rotation curve denoted by V_c (the circular velocity of stars) plotted against the radius of the galaxy. The MOG prediction for the speed of stars fits the data, whereas Newton-Kepler without dark matter does not.**

using dark matter. We wanted to see whether modified gravity alone could account for the stars orbiting at the outer edges of galaxies at about twice the speed predicted by Newton and Einstein. We succeeded in fitting the data for more than 100 galaxies, much of the fitting based on so-called photometric data. These data, acquired from helpful astronomers such as Stacy McGaugh at the University of Maryland, enabled us to obtain remarkable fits to the rotational velocity data with a minimal number of adjustable parameters. In contrast to an adjustable, free parameter, a measured parameter is, for example, Newton's gravitational constant, or the mass-to-light ratio of galaxies. The latter parameter tells us how much mass the galaxy contains, compared to the visible mass, and in the standard model is a telltale sign of the amount of dark matter, if any, in the galaxy.* We were able to explain the data for dwarf galaxies, low surface brightness galaxies, high surface brightness galaxies, elliptical galaxies and giant spiral galaxies, all without

---

* In contrast, astrophysicists adopting dark matter haloes around galaxies use halo models that require two or three parameters as well as the mass of galaxies to fit the rotational curve data. Each galaxy to be fitted requires different values of the two or three dark matter parameters, which results in a large accumulation of free adjustable parameters.

dark matter, because the stronger gravity of the theory was able to account for the speed of the stars. The paper we wrote together, which was published in the *Astrophysical Journal* in 2006, represented one of the most extensive fitting of galaxy data yet achieved by anyone.

Excited by our results, Joel and I immediately set about fitting an equally large amount of data for the mass profiles of hot X-ray clusters of galaxies without dark matter, and succeeded there too, with a minimum number of adjustable parameters. Recall that the clusters also show a significant gravitational anomaly in that there is far too little visible matter in them to produce the gravitational force needed to hold the clusters together, according to Newton and Einstein's theories. Joel and I described our results on the cluster data in a paper published in the *Monthly Notices of the Royal Astronomical Society*, also in 2006.

The fittings to the galaxy rotation Doppler shift data and the mass profiles of the X-ray clusters were achieved without postulating the existence of exotic, nonbaryonic dark matter. The fitting was based on the modified Newtonian acceleration law obtained from either MSTG or STVG, which predicts that the gravitational force is stronger than in Newton's and Einstein's theories.[3] I had combined a modified acceleration law with a varying gravitational constant—part of the STVG theory—to produce an acceleration on a test particle, or a gravitational pull, that increased for a certain well-defined distance range away from the center of a galaxy, and then reduced to Newton's inverse square law of acceleration for large distances from the center of the galaxy.

In other words, what Joel and I were saying in these two papers is that Newton's inverse square law, which describes the effects of gravity over distance and is one of the most hallowed laws of physics, must be wrong, at least on the large scales of galaxies and clusters of galaxies.

MODIFIED NEWTONIAN DYNAMICS (MOND)

One of the unusual features of our fitting the rotational velocity data is that our graphs showing the flattening of the rotational curves plotted against distance from the center of the galaxies were almost indistinguishable from the predictions of the Israeli physicist Mordehai Milgrom's phenomenological MOND formula, even though MOND and STVG/MOG are built on very different underlying principles. Using MOND's empirical formula with its threshold separating Newtonian acceleration from MOND's acceleration, MOND was able, like MOG, to explain the anomalous galaxy rotation curves without dark matter.

An important empirical relationship for galaxies is the Tully-Fisher law, which states that the rotational velocity of a star in a galaxy, raised to a certain power, is proportional to the mass or luminosity of the galaxy. In order to estimate the rotational velocity of the star, we determine from the data at what point the velocity becomes flat, or constant. This contrasts to what Newton's theory demands, namely a diminishing velocity for the rotating stars as one moves from the center to the periphery of the galaxy. Under the Tully-Fisher law, the fixed power of this constant velocity has the empirical value obtained from the data of between three and four. Again, this is in contrast to Newton's acceleration law, where it is the *square* of the rotational velocity of the star that is proportional to the mass or luminosity of the galaxy.

Milgrom's modified acceleration law was designed to reproduce the Tully-Fisher law using the constant power four for the rotational velocity. Joel and I discovered in our fitting routines to galaxy data that the best fits for MOG produced an exponent of the rotational velocity that was closer to three, not four. It turned out that this difference between our fits to the data and those obtained from Milgrom's MOND depends on a subtle interpretation of the data, and at present one is not able to decide which interpretation is better.

Milgrom's MOND model is not a relativistic theory of gravitation. That is, although it successfully does away with dark matter in galaxies, it cannot satisfactorily describe relativistic effects such as the bending of light in a gravitational field or the lensing effects produced by the light from distant objects such as quasars passing near galaxies or clusters on its way to Earth. It is necessary for a modified gravity theory to describe these data without postulating dark matter.

THE VARYING GRAVITATIONAL CONSTANT

One of the salient features of MOG is its varying gravitational constant. As we have seen in the discussion of the variable speed of light cosmology (VSL), varying constants such as Newton's gravitational constant G or the speed of light is not a novel idea. Let us look at the idea of a varying G more closely.

As we have seen, compared to the other forces, the gravitational force is extraordinarily weak—10^{40} times weaker than electromagnetism, in fact. Sir Arthur Eddington was one person who attempted to understand this tiny, dimensionless number, and why it should appear in the laws of physics. He pursued this question from the 1920s until his death in 1944. His fundamental theory, published posthumously by the Cambridge University Press in 1946, was intended to be a unification of quantum theory, special relativity and gravitation.

Eddington first progressed along traditional lines but eventually turned to an almost numerological analysis of the dimensionless ratios of fundamental constants. He combined several fundamental constants to make dimensionless numbers. Several of these numbers would be close to 10^{40}, its square, or its cube root. Eddington was convinced that the mass of the proton and the charge of the electron were natural specifications for constructing the universe, and that their values were not accidental. Researchers stopped taking Eddington's concepts seriously because of his claims about the fine-structure constant alpha, which we recall is a measure of the strength of the electromagnetic forces in atomic physics. Alpha was measured to be very close to one divided by 136, and Eddington argued that it should be *exactly* one divided by 136. Later experimental measurements gave a value closer to 1/137, at which time he then similarly claimed that the value should be *exactly* 1/137. The current measured value of alpha is 1 divided by 137.03599911.

Paul Dirac pursued a similar line of investigation to Eddington's, and called his theory the "large numbers hypothesis." However, in the late 1930s, we recall, he took a new and different approach to gravity. In papers published in *Nature* in 1937 and in the *Proceedings of the Royal Society* in 1938, he described a cosmology with a changing gravitational constant. He reasoned that the tiny dimensionless number 10^{-40} was not just determined by particle physics in terms of the gravitational constant, Planck's constant, the speed of light, and Hubble's constant, but was also determined by the gravitational influence of the whole universe. In order to avoid redesigning all of atomic and nuclear physics, Dirac proposed that Newton's gravitational "constant" G varies with time. This modification of gravity led to a new kind of cosmology in which there was no fundamental significance in the tiny dimensionless ratio 10^{-40}. Dirac postulated that G varies like the inverse age of the universe, so as the universe expanded from the big bang, the gravitational constant, or force, became weaker and weaker as time passed until today, when we experience the present very weak force of gravity. Dirac did not suggest a modification of Einstein's gravity theory to account for his new cosmology, so his theory remained incomplete. However it did yield some predictions, though not very good ones. Dirac's prediction of the age of the universe was approximately 4×10^9 years, or four billion years less than the age of Earth as determined by radioactive dating.

However, as we have seen, Dirac's theory inspired others, such as Pascual Jordan, Carl Brans, and Robert Dicke to take seriously the concept of a varying G.

GRAVITATIONAL LENSING

Although the variation of the gravitational constant in my MOG plays an important role in explaining astrophysical and cosmological data, an equally important role is played by the massive vector phion field. This field creates a repulsive force which, when combined with the varying gravitational constant, leads to a modification of the Newtonian acceleration law and, as we have noted, succeeds in fitting a remarkable amount of astrophysical data. Such a successful fitting to the data cannot be achieved by general relativity without postulating a large amount of dark matter.* MOG is formulated in four spacetime dimensions, so it avoids the problem of string theory or other higher-dimensional theories that are required to hide the extra dimensions from view in our four-dimensional world.

Gravitational lensing occurs when a light ray from a distant source passes by a galaxy or a cluster of galaxies on its way to us. The gravitational field of the massive object acts like a lens, displacing the light from the distant source. Astronomers are using the thousands of galaxies included in the Sloan Digital Sky Survey (SDSS) to investigate gravitational lensing. Data from this extensive survey have shown that general relativity does not fit the lensing data without postulating a large amount of dark matter. Thus, although Einstein's theory was correct on the bending of light in the sun's gravitational field, its predictions on the bending of light by a larger matter source such as a galaxy or a cluster of galaxies are not large enough to fit the observational data.

In my gravity theory, the variation of the gravitational constant and the phion field play key roles in predicting lensing correctly: Due to the increase in the gravitational constant with distance, a light ray suffers a larger bending as it passes a massive source of gravity such as a galaxy. If the galaxy lensing displacement effect occurs close enough to the edge of the galaxy, where the visible stars end, then MOG predicts the shape of the lensing distortion in a unique way that is not seen in Einstein gravity.

THE BULLET CLUSTER

It is difficult to falsify the hypothesis of dark matter, because, as with Ptolemy's epicycles, true believers can always add additional arbitrary features

---

* Neither can the Jordan-Brans-Dicke scalar-tensor-gravity theory fit the data without dark matter.

FIGURE 17: **The famous bullet cluster, said to either prove the existence of dark matter or the correctness of MOG.** Source of *X-ray:* NASA/CXC/CfA/M. Markevitch, et al.; *Lensing Map:* NASA/STScl; ESO WFI; Magellan/U. Arizona/D. Clowe, et al.; *Optical:* NASA/STScl; Magellan/U. Arizona/D. Clowe, et al.

and free parameters to overcome any conceivable difficulties that occur with the dark matter models. However, this is not true of a modified gravity theory, because solutions of the field equations can produce predictions that can be falsified. Hence a modified gravity theory should provide a reasonably tight description of nature that matches the basic properties of any physical system.

In 2004, a group of astronomers at the University of Arizona in Tucson, led by Douglas Clowe, published a paper in the *Astrophysical Journal* in which they described their observations of a cluster of galaxies called 1E0657–56 in the constellation Carina. A series of X-ray observations by the Chandra X-ray Observatory satellite showed that two clusters of galaxies were actually colliding about 150 million years ago, at a red shift of z = 0.296. The remarkable pictures obtained by NASA show a gas cloud offset from the center of the interacting cluster and galaxies on the outer edges of the cluster.

The interpretation is that a smaller cluster of galaxies collided with a larger one. The hot X-ray-emitting gas clouds made of normal baryonic matter in the two clusters were slowed down by a drag force much like a frictional force such as from air, and formed two patches of gas in the center of the colliding clusters. One of these has the shape of a bullet, caused by a bow shock wave from the gases ramming together. Consequently, 1E0657–56 has become known as the "bullet cluster." Since there are huge distances between the galaxies in the merging clusters, they would have passed right through one another, behaving like collisionless bodies, and would be visible to our telescopes at the outer edges of the interacting cluster. One wonders whether observers inhabiting any planets within galaxies of the two clusters would have been aware of the cataclysmic collision.

Two years later, in August 2006, the Tucson astronomers released new data on the electronic archives in the form of a short paper that was later published in *Astrophysical Journal Letters*. A longer paper also appeared on the electronic archives, analyzing the distribution of matter in the colliding clusters and interpreting the gravitational lensing. In the original 2004 paper, the group had claimed that these lensing data from the colliding clusters constituted evidence for the existence of dark matter, although the data were not accurate enough to really substantiate this claim. However, on August 21, 2006, NASA, which had financially supported the project, entered the scene with a dramatic press release. It explained how the new observations magnified the significance of the earlier results with three times as many data, and claimed with more confidence that the astronomers had indeed discovered dark matter. The news of this astonishing "discovery" soon swept around the world in media reports.

From careful measurements of the lensing phenomenon, caused by the bending of light from distant galaxies, one can determine the strength of the gravitational field of a cluster and, in the case of the now-famous bullet cluster, one can also determine the distribution of the matter causing the gravitational pull in the merging clusters. According to the NASA press release, the lensing data tell us that the normal matter associated with the galaxies and the putative dark matter particles in the clusters pass through each other without collision and end up at the outer edges of the interacting cluster. The strength of the lensing of the background galaxy light suggests that there is far more matter in the outer parts of the colliding clusters, where the galaxies are, than the normal baryonic matter of the clusters will allow, according to the lensing predictions of Einstein's gravitational theory. So in order to fit data to theory, the astronomers concluded that there must be at least *ten times* more matter in the outer galaxy regions than was visibly seen—and this then proved the existence of dark matter!

The important point is that the observers claimed that dark matter had separated from the visible hot X-ray-emitting gas in clusters, with which it is assumed to be associated. Normally in a cluster, if one postulates the existence of dark matter, then one cannot separate the dark matter from the normal, baryonic matter in the hot X-ray-emitting gas, which most astronomers believe makes up 90 percent of the visible matter in a cluster. But in the colliding clusters, the interpretation of the astronomers was that as the clusters passed through each other, the X-ray-emitting gas stayed behind in the center, while the visible galaxies *and the dark matter* continued on together. This purported clear separation of dark matter from normal matter led to the claim that dark matter had finally been detected, since in Einstein's gravity theory, without the dark matter near the outside of the clusters, there is not enough gravitational lensing predicted to agree with the data.

The lensing suggested that the dark matter near the outside of the clusters was more concentrated than the matter in the X-ray-emitting gas in the middle. The collision of the clusters had raised the temperature of the X-ray-emitting gas to about twice the temperature of a normal X-ray cluster, making the merging bullet cluster the hottest object ever observed in the universe. Furthermore, the astronomers declared that no alternative gravity theory could explain their cluster lensing data. In particular, they claimed that Milgrom's MOND, the oldest and best known of the current alternatives (although it is merely a phenomenological model, not a theory) could not describe correctly the mass distributions revealed by the lensing data.

These are very dramatic statements, which could have serious implications for the future of physics. If true, they would boost the worldwide efforts to search for the as-yet-undetected particles that could constitute dark matter. Indeed, since August 2006, there has been a groundswell of opinion both in the physics community and in the media that the case is almost closed, that the existence of dark matter has been proved. However, the bullet cluster holds no clue as to exactly what sorts of particles might make up the dark matter within it. More importantly, as we shall see, MOG *can* fit the bullet cluster data without any dark matter at all.

As the NASA press release and the news of the "final detection of dark matter" spread, the physics blog sites on the Internet sizzled with opinions about the purported discovery of dark matter, some enthusiastic, others disparaging. On a popular blog called "Cosmic Variance," Sean Carroll of the University of Chicago and Fermi Lab declared: "And the answer is: there's definitely dark matter there!"[4] In interviews with the media, he went on to say that the existence of dark matter had been proved beyond a reasonable doubt. Moreover, he claimed that it was impossible for a modified gravity theory to fit the bullet cluster data.

Some media reports following the NASA press release also announced that MOND was officially "dead," and on the physics blog sites, in turn, there was much discussion about this pronouncement. One journalist quoted an e-mail from Mordehai Milgrom, who said that the issue was not settled because there was probably a lot of undetected "normal" matter out there in the universe, which I felt was a rather weak response to the powerful new lensing observations.

For several months prior to all this excitement, I had thought about investigating the predictions of lensing in MOG, but had been occupied with other issues. Now I realized that I was forced to study the new observational data and understand how MOG fared. I proceeded to work out the consequences of lensing in my theory, using the same theory, STVG, that Brownstein and I had employed to describe normal mass profiles of hot X-ray clusters. I was excited to discover that the MOG equations predicted an increase in the strength of the lensing in the very regions where the astronomers had identified their dark matter: in the outer lobes of the merging clusters, where the visible galaxies were. The gravitational field caused by MOG's varying gravitational constant, as obtained from the MOG field equations, increased as one moved from the center of the interacting cluster (where the X-ray gas was located) out to the edges, predicting a larger lensing effect in the areas containing the normal visible galaxies, *without any dark matter*. In particular the gravitational pull at the outer edges of the merging clusters where the visible galaxies were situated was significantly stronger than the gravitational field associated with the central gas cloud. This was exactly what the observers were seeing using the Chandra X-ray Observatory satellite. I concluded that with this preliminary result the lensing predictions of MOG were qualitatively consistent with the new bullet cluster observations. I submitted a paper to the electronic archives on August 31, 2006 presenting these results.

The reason that the MOG prediction agrees qualitatively with the data is that the modified gravitational force law in MOG means that gravity increases in strength as you go from the center of a body to the outer part of the body, whether the body is a galaxy or a cluster of galaxies. The effect is minuscule for planets in the solar system; it only becomes significant for large bodies of matter such as galaxies and clusters of galaxies. For clusters of galaxies that are more than half a million parsecs (5×10^{18} kilometers) in size, such as the bullet cluster, the effect can be quite dramatic.

In Newtonian and Einstein gravity, the strength of the gravitational force decreases as you move away from the center of a massive body. But in MOG, gravity *increases* with increasing distance from the center until a certain fixed length scale, determined by the MOG equations, is reached, and then gravity decreases as one would expect in accordance with Newton's law. For clusters, including the bullet cluster, the value of this fixed length scale is about 135 kiloparsecs or 14×10^{17} kilometers. To visualize this, realize that the mean distance between the Earth and the sun is 150 million kilometers; the MOG fixed length scale for clusters is ten billion times larger than that.

One should not draw premature conclusions about the existence of dark matter without careful study of alternative gravity theories and their predictions for galaxy lensing, cluster lensing, and in particular for the interacting bullet cluster. With the correct theory, gravity alone can account for these puzzling data. Joel Brownstein and I have published a longer, more detailed paper in the *Monthly Notices of the Royal Astronomical Society* containing an extensive computer analysis of the bullet cluster data. We obtain excellent quantitative fits to the lensing data from Clowe and his collaborators in MOG without any dark matter.

In addition, MOG has a prediction of the temperature of the larger of the two colliding clusters in the bullet cluster of 15.5 thousand electron volts, or 179 million degrees Celsius. Compare this to the temperature at the core of the sun, which is about 23 million degrees Celsius. The bullet cluster, then, is about ninety times hotter than the core of the sun. This MOG prediction agrees within the errors with the best observations of the bullet cluster temperature. This is a prediction that cannot easily be achieved in a dark matter model.

THE MERGING CLUSTERS ABELL 520

In August 2007, one year after NASA's press release announcing that the bullet cluster had confirmed the existence of dark matter, a paper appeared in the *Astrophysical Journal* by a group of astronomers at the University of Victoria, British Columbia, and Caltech, describing their observations of a cosmic "train wreck" between giant galaxy clusters. At a distance corresponding to 2.4 billion years ago, in the constellation Orion, the merging cluster Abell 520 shows the aftermath of a complicated collision of galaxy clusters. As with the bullet cluster, there are three main components: individual galaxies consisting of billions of stars, hot gas between the galaxies, and the hypothesized dark matter, which can only be inferred through gravitational effects such as lensing. The hot gas was detected by the Chandra X-ray satellite telescope, while optical data from the starlight of individual galaxies were obtained by the Canada-France-Hawaii and Subaru telescopes. The optical telescopes also located the position of most of the matter by means of the light-bending effects from distant galaxies.

The new data deepened the mystery of dark matter for merging clusters because the results contradict those of the bullet cluster. In Abell 520, as with the bullet cluster, there are large separations between the regions where galaxies are most common, and where most of the hot plasma gas is located. However, in contrast to the bullet cluster, the lensing suggests a concentration of dark matter near the bulk of the hot gas, where there are very few galaxies. In addition, there is a region where several galaxies were observed that appeared to have very little dark matter. These observations conflict with the standard ideas about dark matter, and in particular how those ideas are applied to the bullet cluster. As we recall, the prevalent interpretation of the bullet cluster is that the dark matter and the galaxies remain together despite the violent collision. Here, in the Abell 520 cluster, the galaxies and the putative dark matter have separated, and moreover, the dark matter remains with the hot gas.

As the authors of the paper, Andisheh Mahdavi and Hendrik Hoekstra, explained, in interpreting the Abell 520 cluster data, it is necessary to change the basic premise that dark matter particles are collisionless. In other words, there may be two kinds of dark matter: particles that are collisionless, like those used to fit the galaxy rotation data, and particles that *do* interact with their environment and with other dark matter particles. These authors conclude, however, that this interpretation is contrived, as it adds an arbitrary feature to the already mysterious dark matter substance.

Other dark matter proponents simply say that this particular collision of clusters is "messy," and that we have to await further observational confirmation of the data. Still others propose that there is a slingshot effect—that is, the galaxies are shot away from the dark matter and the bulk of the hot gas—but computer simulations of such an effect have not been successful.

How does MOG explain Abell 520? First, we are not constrained to see dark matter together with galaxies, for the obvious reason that there is no dark matter in MOG. Second, the strong lensing that peaks in the bulk of the hot gas and is interpreted as dark matter by the authors, and that has not separated from the hot gas as it should (according to the dark matter paradigm) is likewise not a problem for MOG. As with the bullet cluster, MOG's formula for lensing, which involves a varying gravitational constant G, can explain the multiple lensing peaks in the Abell system with only normal baryon matter.

In MOG, it is possible to obtain a consistent explanation for both merging clusters because of MOG's stronger gravity. In the dark matter model, on the other hand, it is difficult to see how these very different explanations for the two merging clusters can be reconciled without introducing new physics that may then conflict with other observational data such as the galaxy rotation curves.

Ironically, despite the dark-matter hype surrounding the bullet cluster, those data constitute an important milestone in confirming the validity of MOG! The bullet cluster data can be considered one of the most important observational verifications of MOG to date. As for Abell 520, it may turn out that only MOG can provide a satisfactory and consistent explanation of the data.

PART 5

ENVISIONING AND TESTING THE MOG UNIVERSE

Chapter 11

THE PIONEER ANOMALY

T he purpose of NASA's Pioneer mission was to explore deep space within the solar system, out to what was then still called the ninth planet, Pluto. Two spacecraft were launched in 1972 and 1973, identical twins named *Pioneer 10* and *Pioneer 11*. They followed different trajectories, moving from Earth to exit the solar system at opposite sides.

The same unexpected anomalous acceleration or, more properly, *deceleration* has been experienced by both spacecraft. As it has turned out, the Pioneer mission has become the largest scale experiment to date that tests the law of gravity in the solar system.

The two spacecraft were set on trajectories that would in time take them to close-up views of Saturn and Jupiter, and then into the dark cold void of the outer solar system and eventually into the vast empty interstellar space. Heading in the opposite direction to that of the sun's movement through our Milky Way galaxy, *Pioneer 10* will eventually come close to the star Aldebaran, situated some sixty-eight light years away in the constellation Taurus (the bull). It will take about two million years to reach this star. In the meantime, *Pioneer 11* will be heading away from the sun in the opposite direction, and will pass one of the stars in Aquila at about four million AD. Both Pioneer spacecraft carry a gold-plated aluminum plaque designed by Carl Sagan and Frank Drake to be a greeting to other intelligent beings in the universe, if any should happen to intercept the spacecraft. The illustrations on the plaque depict a male and female human being and contain other information about our solar system and the trajectories of the spacecraft.

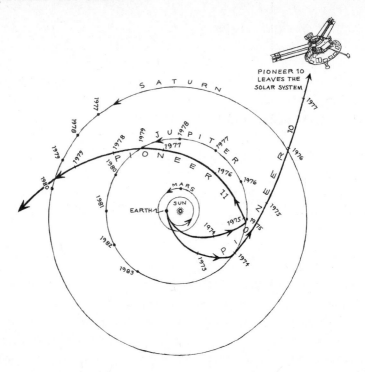

FIGURE 18: The trajectories of *Pioneer 10* and *11* spacecraft from launch to leaving the inner solar system

WHAT IS THE *PIONEER 10* AND *11* ANOMALY?

After launch, as the two spacecraft were propelled into orbits in opposite directions away from the sun, their radio transmitters sent back information to antennas on Earth. From the measured Doppler shift timing data, the precise orbital parameters of the spacecraft were determined. The NASA scientists were surprised to discover in the data an unexpected deceleration of the spacecraft, or what appeared to be a constant, anomalous acceleration toward Earth and the sun. What was showing up was a blue-shifted Doppler frequency of 6×10^{-9} Hertz (Hz), which can be interpreted as a deceleration of the spacecraft as they move away from the sun. After thirty years, this deceleration has resulted in the Pioneer spacecraft both being about 400,000 kilometers off course. This is roughly the distance between the Earth and the moon.

Thus the outward journeys of the space probes were being slowed down by an extra tiny but measurable amount that did not agree with the predictions of Newtonian and Einstein gravity. This effect could have a mechanical explanation. For example, it could be due to a thermal recoil force caused by heat from the electrical equipment in the spacecraft's nuclear power generators or some other unknown mechanical effect. On the other hand, it could be due to a new physical phenomenon such as a force of gravity stronger than is predicted by Newtonian and Einstein gravity theories.

In November 2005, I was invited to a special meeting in Bern, Switzerland, sponsored by NASA and the International Space Science Institute (ISSI). It was a select meeting in the beautiful old city where Einstein had worked in the patent office and written his five famous papers in his miracle year of 1905. Only about fifteen scientists, both experimentalists and theorists from France, Norway, Germany, Portugal, Canada and the United States, had been invited to the meeting. The purpose of the workshop was to try to get to the bottom of the anomalous Pioneer data, and to figure out what was causing the unexpected effect.

The convener of the meeting was Slava Turyshev, a charming, twinkly-eyed Russian-born physicist from the Jet Propulsion Laboratory (JPL) in California, a division of NASA. He was one of the authors of the scientific paper first reporting on the anomalous acceleration of *Pioneer 10* and *11* in *Physical Review Letters* in 1998.* It was obvious from his introductory remarks at the start of the three-day meeting that Slava possessed an inexhaustible capacity for technical details and an indefatigable constitution for describing enormous amounts of technical information. He was the ideal person to head up the data analysis and to chair the workshop.

To the amazement of most of us in the room, Slava explained in his formal talk how the JPL team had recently discovered boxes of old tapes with recorded data from the Pioneer launches in 1972 and 1973. These contained accurate data from the time of the spacecraft launches to the position of Saturn and beyond. Remarkably, someone had found the boxes of old tapes just a week before JPL was to throw them, along with other old trash, into a Dumpster. These data have now been digitalized for use in modern computers, and teams in France, Germany, and at JPL are analyzing them. Slava said he hoped that the analysis would be completed within a year after the Bern meeting.

---

* Another author on the paper was the senior JPL astronomer John Anderson, who had years of experience in analyzing deep space probe data. Michael Nieto, a theorist at Los Alamos National Laboratory, New Mexico, had lent his theoretical knowledge of gravity experiments to the project.

COULD THE PIONEER ANOMALY HAVE
A MECHANICAL EXPLANATION?

The first two days of the workshop were devoted to talks attempting to pin-point any mysterious systematic but conventional effect that could be caus-ing the anomalous acceleration. The sun produces a solar wind, so could its pressure be influencing the motion of the Pioneer spacecraft by small amounts? A straightforward calculation shows that this solar wind pressure drops off rapidly as the spacecraft move away from the sun, and therefore this can not be the cause of the anomalous acceleration.

What about a possible internal cause of the anomaly that could have something to do with the spacecraft itself, like electrical wires or out-gassing of propellant (fuel leaks)? The NASA team and their collaborators had in 2000 published an exhaustive review in *Physical Review D* of all the possible physical effects that could create a systematic anomalous acceleration, and they came up with a negative answer: There seemed to be no obvious aspects of the spacecraft design that could cause the observed effect.*

Other external effects such as solar radiation pressure, drag forces due to interplanetary gas and dust, and the gravitational effects of the planets and Kuiper belt objects were also considered as possible sources of the anomaly, and they, too, were ruled out by the investigators.

The big issue at the meeting was how to model the properties of the Pioneer anomaly, in order to elucidate the origins of the anomalous accelera-tion. Unfortunately, there are no absolute measurements of the positions, speed, and acceleration of the spacecraft to determine these physical quanti-ties. Therefore, experimentalists use Doppler data to interpret the dynamical orbital information about the spacecraft with respect to Earth. In addition to the Doppler data, the spacecraft transmit telemetry information, which pro-vides information about voltages, temperatures, currents, and orbital thrust duration times. All of these sources of data could point to various mechanical changes suffered by the spacecraft during their long voyages.

Viktor Toth, a clever and skillful Hungarian-Canadian physicist from Ottawa who is Slava's collaborator, gave a presentation about the engineer-ing structure of the Pioneer spacecraft. A software engineer and computer expert and the only other Canadian participant at the meeting, Viktor has collaborated with the Pioneer team for several years. His expertise lies in understanding the spacecraft design and analyzing the flight mission,

---

* This conclusion proved to be premature; today the investigators are still searching for a mechanical or heat effect.

searching for possible internal and external causes of the anomalous acceleration. At one point during his talk, Viktor said, "It's important for us to consider the three-meter antenna dish on the spacecraft." Slava immediately jumped up, waving a hand, and interrupted. "Viktor, it's a 2.74-meter dish. We're not allowed to round off numbers here!" It was clear that Slava was not putting up with any imprecisions. Paying attention to details is the only way that we can hope to answer the questions raised by the Pioneer spacecraft.

Slava stressed the importance of analyzing every internal aspect of the spacecraft designs. He was particularly concerned with the power supply, which comes from four onboard generators. This heating could conceivably produce very small thrusts opposite to the direction of radiation. If it were found that this heat radiation was in the same direction as the Pioneers' trajectory, this could slow down the spacecraft and account for the anomaly.

In other words, there are multiple possible causes for the tiny acceleration observed. Slava emphasized the fact that as observationalists, he and his colleagues had to maintain an objective attitude toward the whole project: The investigation of the anomalous acceleration should not be swayed by any theoretical explanations involving "new physics." He said, "We've got to keep an objective point of view and continue to investigate any possible spacecraft design origin of the anomalous acceleration."

CAN A MODIFIED GRAVITY THEORY EXPLAIN THE PIONEER ANOMALY?

On the third day of the meeting, it was the theorists' turn to present their ideas on any possible new physics interpretations of the data. Two physicists who had analyzed the observational data with computer codes, taking into account the orbital parameters of the spacecraft, both concluded that the anomalous acceleration was definitely in the data; it was not some kind of artifact of the measurements. This meant that three independent groups had now confirmed that the anomalous acceleration is indeed happening, which is a very important result because it means that we are not pursuing some phantom effect that is not truly there.

A German physicist proposed that undetected dark matter in the solar system might explain the anomalous effect. However, he was not happy with the results of his calculations. In addition, the Portuguese physicist Orfeu Bertolami and I pointed out that the amount of dark matter that would be needed to produce the anomalous acceleration would be far too large compared to the density of putative dark matter in galaxies and in large-scale cosmology. The bounds on dark matter in the solar system would not permit the kind of dark matter density that he required. As 'Orfeu said, "With this kind of dark matter density, you would destroy the solar system!"

Serge Reynaud from the University of Paris, who headed the theory group talks, presented his modification of Einstein gravity, which he has published in collaboration with Marc Thierry Jaekel from the same institution. Reynaud considers the Pioneer missions a large-scale gravity test. According to him, it is possible to modify Einstein's general relativity for the outer solar system without conflicting with the very accurate gravity tests performed on Earth and for the inner planets. Reynaud and Jaekel claim that a special kind of mathematical modification of Einstein's gravitational theory, without introducing new fields, could explain a new effect associated with the Doppler shift and time delay of radio signals from the spacecraft. This effect would cause the anomalous acceleration observed. Reynaud claimed that measurements of the speed of a spacecraft in a future mission to the outer planets could verify their theory. In particular, the Pioneer anomalous acceleration would depend upon the square of the velocity of the spacecraft with respect to the sun.

In my talk that afternoon, I reviewed my Modified Gravity theory (MOG) and explained how it fit the rotational velocity data for galaxies and also the data for X-ray clusters well, without dark matter. Then I described the application of my theory to the Pioneer anomaly, which I had published recently in collaboration with Joel Brownstein in *Classical and Quantum Gravity*. We fitted the anomalous acceleration data published by the JPL team and discovered, using MOG with two free parameters, that the fit was far better than we had anticipated. JPL's published acceleration data before 1983 had large errors. However, the central values of the data within their error bars produced this excellent fit, from which we concluded that the central values were far more reliable than had been expected, and should be taken seriously. The important result of our analysis was that the anomalous acceleration began to appear in the data when both spacecraft reached the orbit of Saturn.*

One vital piece of information needed to permit a gravitational interpretation of the anomaly is to know where the anomalous acceleration is pointed—toward the sun or toward the Earth. If the anomalous acceleration vector is pointed toward the sun, this tells us that the origin of the anomaly could be gravitational, for the sun produces by far the largest gravitational

---

* The JPL acceleration data must be verified by the analysis of the data on the old tapes, which is still being carried out.

field in the solar system. If the anomalous acceleration is pointed toward the Earth, however, this would remove the possibility that gravity is causing the anomalous acceleration, because the Earth's gravitational field is obviously not strong enough to produce the acceleration; it would open the door to a mechanical or heat effect as the cause of the anomaly.

An important side effect of all this, I explained in my talk, is that according to my gravity theory, the anomalous Pioneer acceleration is either zero or so small as to be undetectable between the orbits of Earth and Saturn. Therefore my theory agrees with all the accurate *inner solar system data*, which do not show a fifth force component to gravity or a violation of Einstein's equivalence principle. Recall that in MOG both the fifth force and the variation of the strength of the gravitational constant G combine to strengthen gravity, yet MOG's fifth force becomes vanishingly small for the inner planets, including Earth. So the Pioneer data until the orbit of Saturn agree with the predictions of Einstein's and Newton's gravity theories, as well as with MOG.

I then established the fact that Einstein's and Newton's gravity theories have not been well tested for the outer planets beginning with Saturn. Observational data for the outer planets today are not accurate enough to exclude a modified gravitational explanation for the anomalous acceleration. This seemed not to have been understood previously. Indeed, the tiny size of the anomalous acceleration—about one ten-thousandth times smaller than the Newtonian acceleration experienced by the spacecraft—demands highly accurate measurements that have not as yet been performed.

Although no consensus was reached at the meeting on the cause of the Pioneer anomaly, the possibility of analyzing new data from the old files gave us hope for settling many of the important issues in the future. Slava's prediction that the newly discovered data would be analyzed quickly turned out to be overly optimistic. Three years later, the analysis still continues in France, Germany, Canada, and California. And there are now yearly workshops to discuss progress.

EINSTEIN'S BERN

Holding the workshop in Bern meant that we were aware of the spirit of Einstein hovering over our discussions, especially since this was the centennial celebration of Einstein's five important papers in 1905, and the local museum had a special exhibition on his life. I took some time on my own during our few days there to walk the cobblestone streets that Einstein had walked in 1905. I crossed the Kirchenfeldbrücke over the dizzyingly deep gorge through which the River Aare flowed—the same bridge that young Albert crossed daily between his home at Kramgasse 49 in the old city and the patent office on the south side of the river. I imagined him strolling along the bridge, his hands in his pockets and his mind far away, chasing light beams.

Chapter 12

MOG AS A PREDICTIVE THEORY

One way of judging the value of a fundamental theory of physics is to ask: How much data can be fitted with a minimum number of assumptions and free parameters? Measured constants like the Newtonian gravitational constant and the current speed of light are not free parameters. Moreover, the mass of a body such as the sun or a star or a galaxy can be determined by observation, and consequently this is not a free parameter. A free parameter is a number that can be adjusted in an ad hoc way to fit some given observational data. It can be considered a fudge factor.

Dark matter models have many such free parameters that are adjusted to fit astrophysical data such as the rotational velocity curves of galaxies. However, dark matter is not strictly speaking a fundamental theory. It is simply incorporated into Einstein's gravity theory so that the theory can fit astrophysical and cosmological data.

The modified Newtonian law of gravity obtained from the MOG field equations contained two free parameters. One related to the strength of the coupling of the phion vector field to matter. The other was the distance range over which the modified acceleration law acts. The earlier work on MOG—fitting galaxy rotation curve data, X-ray clusters, cosmological data, and the Pioneer anomaly—required adjusting these two parameters to each physical system being considered. Although the published fits to astrophysical data were remarkably good, we still had to face the criticism that we had used two adjustable free parameters to obtain the results.

BANISHING THE FUDGE FACTORS

Ideally, when we solve the field equations of MOG, the solutions should *determine* the actual values of any parameters in the theory and how they vary with space and time as we go from the solar system out to the horizon of the universe. Recently, during the writing of this book, I began a profitable collaboration with Viktor Toth, whom I had met at the Pioneer anomaly workshop in Bern. Viktor had shown interest in learning about how MOG works as a theory of gravitation, and why the fits to the astrophysical data that Joel Brownstein and I had published were so good. Viktor and I began our collaboration on MOG by attempting to fit the theory without dark matter to globular cluster data and data obtained from the Sloan Digital Sky Survey (SDSS) galaxy survey that applied to satellite galaxies.* Next we tackled the more difficult problem of deriving a full explanation through MOG of the large amount of published cosmological data from Wilkinson Microwave Anisotropy Probe (WMAP) and the large-scale galaxy surveys. This research proceeded well, and we soon found that MOG could fit all the presently available data remarkably well with a minimum number of free parameters.

It then became evident to us that when we solved the field equations for the scalar fields that determined the variation of the coupling strength of the phion vector field to matter, the effective mass of the phion field or its distance range, and the variation of the Newtonian gravitational coupling constant, then MOG could actually *tell us what the values of these parameters were*, so they were no longer free parameters.[1] MOG now had no free parameters at all!

This was a thrilling result because it meant that MOG had now become a truly predictive theory. Further research disclosed that the solutions of the field equations were so powerful that given the mass of a physical system and the Newtonian gravitational constant, MOG could, without any free parameters, predict observational data coming from our planet Earth and the solar system, to globular clusters the size of a few parsecs, to galaxies and clusters of galaxies, and, finally, to all the cosmological data—all without any dark matter. We had arrived at the dramatic result that MOG could predict the effects of gravity over fourteen orders of magnitude in distance scale, from the short distances on Earth out to the farthest horizon of the universe without any free parameters or dark matter. This is a predictive power comparable to Einstein's application of general relativity to the solar system in 1916 with his prediction of the perihelion advance of Mercury

---

* Globular clusters, to be discussed in Chapter 13, are relatively small, dense conglomerations of up to a million or more stars occurring commonly in galaxies. Satellite galaxies, such as the Large and Small Magellanic Clouds, orbit host galaxies or even clusters of galaxies.

and the bending of light. This was an emotionally exciting time, as Viktor and I realized that we had now arrived at the home stretch in developing the theory. Could this mean that we had demonstrated that MOG was a truly fundamental theory of nature?

BEGINNING TO MAKE PREDICTIONS: THE SOLAR SYSTEM

Armed with the new predictive power of MOG, we were able to establish that the modified acceleration law in the theory reduces to that of Einstein and Newton within the solar system. This guarantees that MOG agrees with all the very accurate inner solar system data, including the Eötvös experiments testing the equivalence principle on Earth, the very accurate lunar laser ranging experiments, the radar ranging experiments for Mars, and the Cassini satellite space mission data for Saturn. Thus MOG is now predicting that Einstein's general relativity agrees with considerable accuracy with all solar system tests, including the bending of light by the sun.

REVISITING THE PIONEER ANOMALY

The previous chapter on the Pioneer anomaly was written not long after the 2005 Bern workshop. Much has happened in the development of MOG since then. Building a new theory in physics means overcoming unforeseen obstacles, and wandering down research avenues that may later turn out to be dead ends, or perhaps hidden passageways into new, more promising avenues. Today we have a much more powerful theory than in 2005, with no free parameters. When we apply this mature MOG to the solar system, we find that it does not, after all, support the hypothesis that the anomalous acceleration of the Pioneer spacecraft is due to the sun's gravity. Contrary to what I believed at the time, the predictive MOG actually suggests that the Pioneers' anomalous acceleration is a *nongravitational effect*, such as a thermal or out-gassing effect—for there appears to be no other effect large enough to account for the anomaly.

When we calculate the acceleration for Pioneer 10 using MOG's fifth force correction to Newton's law of gravity, we arrive at a tiny number: less than 10^{-22} meters per second squared. This is the maximum anomalous acceleration due to the sun's gravity that MOG predicts should be experienced by Pioneer 10 by the time it reaches Pluto's orbit. In fact, this acceleration is many orders of magnitude smaller than the observed anomalous acceleration of both Pioneer spacecraft. This implies—if MOG is correct—that something besides gravity is causing most of the acceleration. The most likely cause for an effect of that magnitude would be a thermal effect originating on the spacecraft themselves.

In the previous chapter on the Pioneer anomaly, we saw that at that time, MOG did not possess the predictive power it now does, and we were required to fit the free parameters in the MOG calculation of the anomalous acceleration to the Pioneer anomaly data. Now, however, since the theory has no adjustable free parameters, we can use it to make a genuine prediction, rather than merely fitting the theory to the data. MOG's new prediction of a nongravitational effect to explain the Pioneer anomaly may be confirmed through more rigorous data analysis by JPL and their collaborators during the next year. This would in turn confirm that Einstein and Newtonian gravity agree with the solar system data down to the tiny level of an acceleration of 10^{-22} meters per second squared out to the orbit of Pluto.

MOG, of course, contains Newton's and Einstein's theories of gravity, just as Einstein contains Newton. This new MOG prediction for the Pioneer data suggests that MOG may only come into effect in gravitational fields of massive bodies such as galaxies—unless someday a space mission is able to detect such minuscule gravitational effects in the solar system as MOG's tiny difference for gravity, compared to Einstein and Newton.

THE BINARY PULSAR PSR 1913+16

MOG has made very accurate predictions for the orbits of two neutron stars in the binary system PSR 1913+16. In 1974, using the Arecibo 305-meter antenna telescope, Russell Hulse and Joseph Taylor discovered in the Milky Way galaxy 21,000 light years away from Earth the first known binary pulsar system, a discovery that earned them the Nobel Prize in 1993. A pulsar is a neutron star that rotates very rapidly and produces a pulsing signal of light, much like a rotating lighthouse. The analysis of these binary data led to the first indirect detection of gravitational waves.

The reason that MOG actually predicts the binary pulsar data is that the distance between the two binary stars is approximately the distance between the sun and the inner planets, which means that Einstein's gravitational theory holds to a high degree of accuracy for this system. As we recall, the strength of gravity in MOG is dependent upon distance in a counterintuitive way: The more distant a body is from a gravitating source of matter, the stronger is the gravitational force acting on the body. This is a surprising feature of MOG, in contrast to Newton's inverse square law, which states that gravity decreases with the square of the distance, as an object moves away from a gravitational force. In MOG, this increase of gravitation with distance holds true until a certain fixed distance, which depends upon the size of the gravitational system. Beyond this fixed distance, MOG's force law becomes the same as Newton's inverse square law. Because of the closeness of the stars in this binary pulsar system, the increase in the strength of MOG's gravity does not yet show itself, just as in the solar system. Thus, this binary pulsar system does not constitute a test that can distinguish between MOG and Einstein's gravity theory. It merely shows that MOG and Einstein's theory are the same at relatively short distances.

GLOBULAR CLUSTERS: ANOTHER TEST OF MOG?

Studying globular clusters, however, *does* provide an important test that can distinguish among gravity theories such as MOG, Milgrom's MOND, Bekenstein's TeVeS, and Einstein and Newtonian gravity. Most astrophysicists consider globular clusters—dense groups of stars found quite commonly in galaxies—to be among the oldest objects in the universe, although others claim that they can also be formed in the later universe when galaxies merge and their stars collide. Such globular clusters will be younger and will have a higher metal content than the older ones.

Observations show that the mass-to-light ratios of the globular clusters are about one. That is, the luminosity of the globular cluster, which you can see, is about the same as the amount of measured visible matter. This, plus the fact that the random motion of stars in the globular clusters fits Newtonian gravity, leads astronomers to conclude that there is no dark matter in globular clusters. Globular clusters can be situated close to the center of the galaxy or in its outer edges. The globular clusters in the outer parts of our Milky Way galaxy are not as subject to strong gravitational forces from the center of the galaxy, so they make ideal astronomical objects to distinguish between dark matter and modified gravity. Since external tidal forces due to the strong gravity of the galactic center do not disrupt them, we can consider them as isolated gravitational systems.

Milgrom's MOND states that if the acceleration experienced by a body is greater than a critical acceleration a_0, then Newtonian dynamics prevails. On the other hand, if the acceleration is smaller than the critical acceleration a_0, then the body is said to be deep inside the MOND regime, and Newton's law of gravity is modified. We know from the fits to galaxy rotation curves that the critical acceleration in MOND is 1.2×10^{-8} centimeters per second squared. Those globular clusters that are closest to the galactic center have an acceleration greater than this critical value of a_0 and therefore Newton's law of gravity is not modified. Those in the outer edges of the galaxy, on the other hand, have an acceleration smaller than the critical acceleration. Moreover, the internal acceleration experienced by the stars inside the globular cluster is also smaller than a_0. Therefore the acceleration for these outlying globular clusters is deep inside the MOND regime and MOND's predicted velocities of the stars in them should agree with the observational data.

Astronomy groups at Sternwarte, University of Bonn, Germany, and the Astronomical Institute of the University of Basel, Switzerland, have recently performed observations with a ten-meter telescope of the velocities of the stars in the outlying globular clusters. They find that these velocities match those predicted by Newtonian gravity—without having to add in any dark matter—and do not agree with the prediction of MOND. In particular, a globular cluster called PAL14—with a small mass of about 10,000 times the mass of the sun and a rather large radius of 25 parsecs or 75 light years—has a mass-to-light ratio of two. Hence the consensus is that there is little or no dark matter in PAL14. The discrepancy with the MOND prediction is significant. MOND's predicted mass-to-light ratio is far too small to agree with the observational data. Now we appear to have an astronomical system in addition to the bullet cluster that rules out MOND as a possible explanation of the dynamics of galaxies and clusters of stars. Provided continued observations of these outlying globular clusters support these published data, then MOND, and in turn TeVeS, will be ruled out as a modified gravity theory.

Viktor Toth and I have recently predicted the dispersion velocities for PAL14. Dispersion velocities are the speeds of randomly moving stars in a stable system such as the globular cluster. By using the MOG acceleration law, we obtain excellent agreement with the velocities observed for the globular cluster PAL14. The predictions for the velocities of the stars in the globular clusters are the same as the Newtonian predictions because of the small distance scales of globular clusters compared to galaxies. This confirms that MOG fits the globular cluster data without dark matter as well as it fits the galaxy rotation curve data, and the data for the clusters of galaxies. The fact that Newtonian gravity also fits the globular cluster data without dark matter does not mean that Einstein/Newtonian gravity is correct, because we know that the standard theory does not fit the galaxy rotation curves, nor does it fit the clusters of galaxies data or the lensing data for galaxies and clusters. All of those cases need either dark matter or MOG. Thus the globular cluster data could be just as compelling in proving that MOG is correct as the bullet cluster is. The bigger the system, and the farther away you are from its center, the more MOG with its distance-dependent G comes into play.

How do dark matter proponents get around this problem? Some have speculated that during the long period of evolution of the older globular clusters, over ten billion years, the dark matter that originally existed in the clusters at the beginning of their evolution has somehow been stripped away. Another speculation is that globular clusters may form from the cores of dwarf galaxies found within larger galaxies. Dwarf galaxies are larger than globular clusters and smaller than spiral and elliptical galaxies. Yet the consensus of astronomers is that dark matter does not occur in the cores of galaxies, but only as haloes around galaxies. And so, this explanation runs into more trouble—why is there no dark matter in the cores of galaxies? In fact, some computer models of dark matter insist that there should be a great deal of dark matter in the cores of galaxies, disagreeing with most interpretations of astrophysical observations! This has been a flaw in the dark matter scenario for years.

I do not find any of these explanations for the lack of dark matter in globular clusters and in the cores of galaxies convincing. In contrast, the MOG field equations predict simply that the gravity in the cores of galaxies conforms to Newtonian gravity, and is due only to visible baryonic matter.

GALAXIES AND CLUSTERS OF GALAXIES

When I first began developing the modified version of Newton's law of gravity in 1995, I discovered in collaboration with Igor Sokolov that in order to fit the galaxy dynamical data without dark matter, it was necessary that the modification of Newtonian acceleration be proportional to the square root of the mass of the galaxy being considered. This ratio in fact fitted all the galaxy rotation data very well. However, some physicists and astronomers complained that this feature of the modified gravity law did not appear from the basic principles of MOG in a natural way, but seemed to be a rather ad hoc derivation. The reason for this troubling aspect of the modification was that Igor and I had to fit the Tully-Fisher law. We recall that this relation for spiral galaxies required that the fourth power of the rotational velocity curves of stars in the galaxies be proportional to the mass of the galaxies. In Newtonian gravity this is not true; rather, it is the second power, or square, of the rotational velocity that is proportional to the mass.

However, Viktor and I now arrived at the startling result that when we solved the field equations of MOG, the solutions provided this information that fitted the Tully-Fisher law and gave an excellent fit to the rotational velocity data of the galaxies. All of a sudden, all the observational data results clicked together like a Rubik's Cube snapping into place. We now had a fundamental explanation for the data without ad hoc assumptions.

During the calculations needed to resolve these issues, one or two mistakes were made, and when we discovered these mistakes and corrected them, the fits to the data improved. It was as if MOG was telling us: "Don't do that. *This* is what I need to make me happy!" In theoretical physics research this is often not the case. Usually, when a mistake is made and is corrected, everything falls apart. And so, as Viktor and I proceeded with our calculations, we began to feel that we were indeed on the right path to a truly important discovery revealed by nature through this mathematical theory.

Applying MOG's modified acceleration law to galaxies and clusters of galaxies in its new predictive form led to a remarkable discovery. Given the mass or, equivalently, the mass-to-light ratio, of a galaxy, beginning with the small dwarf galaxies that have a radius of about 8,000 to 10,000 parsecs and passing on to giant spiral galaxies with masses a hundred times larger and extending over a radius of 70,000 to 100,000 parsecs, we discovered that MOG could in fact predict the rotational velocity curve data without adjustable free parameters and—of course—without dark matter. We recall that previously, Joel Brownstein and I had fitted the galaxy rotation curve data for 100 galaxies using two adjustable parameters that changed going from dwarf galaxies to the giant spiral galaxies. Now, we were able not only to fit the data, but actually predict the speed of the stars in all the known types of galaxies without any extra parameters! Conversely, if we knew the speed of the stars in their orbits around the galaxies, we could predict the mass-to-light ratio of each galaxy, again without any extra parameters and—of course—without any dark matter. The basic equations of MOG were now superior to any other known modification of gravity theory, including Milgrom's MOND, which has one adjustable parameter, and current dark matter models, which require two or three free parameters in order to fit the rotational velocity curves for each different galaxy.

In a similar fashion, Viktor and I now applied MOG to the X-ray mass profiles of clusters of galaxies, and again without any adjustable parameters or dark matter, given the mass of a cluster, we were able to predict its stability. That is, we were able to predict the stability of galaxy clusters based on MOG's stronger self-gravitating forces acting between the galaxies in the cluster. With the new predictive power of MOG we can also explain the bullet cluster data even more convincingly than before, when we had used two adjustable parameters in the MOG acceleration law.

SATELLITE GALAXIES

From the SDSS data, which surveys millions of galaxies in the sky, astronomers have been able to identify satellite galaxies that orbit host galaxies or clusters of galaxies. Such data can provide another test of gravity theories, because the host galaxies can be treated as central masses binding satellite galaxies to them through gravity. The astronomers Francisco Prada and Anatoly Klypin have analyzed satellite galaxy velocity data using the N-body simulation computer code and find that they can explain the observed speeds in the orbits of satellite galaxies using the standard Lambda CDM model, which assumes that at least 90 percent of the host galaxy masses consists of dark matter halos. They also claim that they cannot fit MOND to these data without using dark matter. However, MOND theorists such as HongSheng Zhao and his collaborators at the University of St. Andrews in Scotland make certain assumptions about the dynamics of the satellite galaxies that allow them to fit MOND to the data without dark matter.

How does MOG fare with the satellite galaxy data? Viktor and I found that the parameter-free MOG theory explained the satellite galaxy data well without dark matter: Using the MOG equations, no dark matter is needed in the host galaxy to keep the satellite galaxies in orbit. Unfortunately, the satellite galaxy data do not constitute a definitive test at the present time to distinguish between the Lambda CDM model, MOND, and MOG, as all appear to fit the data equally well. Perhaps more accurate data for the satellite galaxies will ultimately allow us to distinguish one theory from another. In the future Viktor and I plan to redo our calculations for the satellite galaxies using an N-body simulated computer code based on the MOG field equations.

THE NEW LAW OF INERTIA

Viktor and I embarked upon an attempt to understand how Mach's principle, which says that the inertia of a body is produced by the influence of all the matter in the universe, would be understood in MOG. In Einstein's gravity theory it is not possible to properly invoke Mach's principle. The origin of the inertia of a body has been a great mystery ever since Newton's bucket experiment and his discovery of his theory of gravity. In 1953 Dennis Sciama, who brought me to Cambridge and matriculated me at Trinity College, published a paper in the *Proceedings of the Royal Society* attempting to explain

Mach's principle in gravitational theory. Although his ideas were brilliant, ultimately he did not succeed.

Newton proved in his theory of gravity for the motion of the moon that the gravitational field outside a spherically symmetric body behaves as if the whole mass of the body were concentrated at the center. This is a difficult concept to understand without mathematics, and in fact it took Newton twenty years to prove it. This result also applies to Einstein's gravity theory. According to a theorem proved by George Birkhoff, a mathematician at Harvard University who published a book on relativity in 1923, a spherically symmetric gravitational field in empty space in Einstein's gravity theory must be the static metric given by the Schwarzschild solution. The theorem also states that the gravitational field inside an empty spherical cavity of a spherically symmetric body is zero. This is analogous to the result in Newtonian gravity where the gravitational field of a spherical shell vanishes inside the shell.

On the other hand, in MOG, only a particle located exactly at the center of a hollow shell does not, because of reasons of symmetry, experience a force. However, when the particle is deflected from the center, it will experience a force that, as Viktor and I calculated, would try to push the particle back to the center.

We live inside a universe that is filled with approximately homogeneous matter. We can imagine dividing up the universe into a series of concentric shells centered around a particle. When a force acts on the particle and moves it away from the center of these concentric shells, MOG produces a force equal in magnitude but opposite in direction, pulling the particle back. This is precisely how the law of inertia works, leading us to conclude that this counteracting force is the inertial force.

Think of pushing against a chair with your foot. You feel that the chair is resisting the force of your foot. In Newton's mechanics, this resistance is referred to as the inertial mass of a body. But it can also be thought of as "inertial force." Mach postulated that the inertial force of a chair resisting your foot—like the concave surface of water in Newton's rotating bucket—was caused by all the distant matter in the universe. If we picture the chair sitting in an empty universe without any other matter present—except for *you*—then when you push the chair with your foot, there would be no force of inertia resisting your push and at the slightest touch the chair would just fly away into space with extreme speed. In MOG the law of inertia evolves naturally from the theory's mathematics. MOG's cosmological field equations determine the average density of matter in the universe, which then creates the inertial force. All of this is consistent with Newton's second law, namely, that force equals mass times acceleration, where mass is identified as the inertial mass of a body.

This result was actually postulated to be true a long time ago, by the French mathematician Jean le Rond d'Alembert in 1773, and it is known as D'Alembert's principle of dynamics. He invented an inertial force, but he had no theory to explain how it came about. But his ideas contained the seed of the understanding of the inertia of a body, which we explained more fully through MOG.

Our calculations demonstrated that the inertia of a body followed from the MOG law of acceleration. But more dramatically, we discovered that when the acceleration of a body was very small, about 10^{-10} meters per second squared, then the equivalence of gravitational mass and inertial mass, which played a profound role in the development of Einstein's theory of gravity and was assumed to be true in Newtonian gravity, was violated by a small amount.* This exciting discovery led us to propose an experiment to test this prediction of MOG: In a freely falling system such as the environment in the international space station, one could apply a tiny electric force to a test particle, simulating an acceleration of the particle of about 10^{-10} meters per second squared, and check to see whether Newton's law of gravity held true. We predicted that it would not hold true because according to MOG, the inertial mass of the test particle was not quite equal to its gravitational mass. Hopefully this experiment in space can be performed in the not-too-distant future.

Our recent work has opened the door to discovering new predictions coming from MOG, which could distinguish between this modified gravity theory without dark matter and Einstein gravity with dark matter. A truly fundamental theory, as I have said, should be able to make predictions that can be verified by observational data or tested by experiment, and that cannot be explained equally well by any other competing theory. There are few examples of such fundamental paradigms in physics. One outstanding example is Maxwell's wave solutions of his electromagnetic field equations that predicted that electromagnetic waves traveled at the speed of light in a vacuum. Another example is Paul Dirac's solution of his quantum equation for the electron that predicted the existence of a particle with the same mass as the electron but with the opposite charge. This prediction was confirmed

---

* This acceleration was remarkably close to the value of the Pioneer anomaly and also close to MOND's critical acceleration—which may only be a coincidence, without profound implications.

by experiments two or three years after the publication of the famous Dirac equation, when the positron was found.

As we have learned, the new predictive power of MOG allows us to make specific predictions in cosmology and for astrophysical phenomena such as binary pulsars, globular clusters, galaxies, clusters of galaxies, and satellite galaxies.

Chapter 13

COSMOLOGY WITHOUT DARK MATTER

A major goal of MOG is to remove the need to postulate dark matter—to be able to describe correctly the universe that we can actually see, detect, and measure. This becomes a particularly acute problem in cosmology, the study of the large-scale structure of the universe. A considerable amount of research has been published developing the standard model of cosmology using dark matter. A MOG has to compete with this huge amount of published research and come up with satisfactory explanations for the wealth of observational data obtained during the last two decades in cosmology. This is a daunting task.

FLIRTING WITH DARK MATTER

In a 2006 paper exploring modified gravity, I attempted to use my MOG's phion vector field and its massive vector particle as a stand-in for dark matter. The phion particle is a boson (like the photon and mesons) that carries MOG's fifth force. It has spin 1 times h, where h is Planck's constant.

It is known in laboratory experiments that at very low temperatures, electrically neutral bosons can form a new kind of matter called a "Bose-Einstein condensate." This phase of matter was predicted by Einstein in 1925 on the basis of work published by the Indian physicist Satyendra Nath Bose in 1924, in which he introduced a new kind of quantum statistics associated with boson particles. These statistics describe the behavior of a large number of bosons in a gas. They are in contrast to Fermi-Dirac statistics, discovered independently by Enrico Fermi and Paul Dirac in 1926, which describe a gas of fermions, such as electrons and protons. Unlike fermions, bosons love to get together and form a large crowd of particles, and in particular when photons crowd together, they can form a coherent system of particles known as a laser beam.

The Bose-Einstein condensate is a unique form of matter. It has the interesting property of having negligible pressure and viscosity. From a theoretical point of view, the Bose-Einstein fluid is formed through a phase transition, analogous to the condensation of water from steam. In physics we talk about a phase transition occurring through spontaneous symmetry breaking, such as when iron filings suddenly become magnetized and rearrange themselves in an aligned pattern in an external magnetic field. This spontaneous symmetry breaking, first described at the atomic level by the celebrated Nobel laureate Werner Heisenberg, is a purely quantum mechanical phenomenon. Thus the formation of a Bose-Einstein fluid is explained by quantum mechanics, not classical physics.

In my 2006 paper, I suggested that matter in the early universe formed as a Bose-Einstein condensate fluid from the phion boson particles sticking to one another. Remnants of this condensate would still be found in galaxies today. Identifying these phion Bose-Einstein condensates with the dark matter provided me with the much-needed 26 percent of the missing matter. When combined with the 4 percent of visible baryonic matter, this constituted the 30 percent of matter needed to fit the Wilkinson Microwave Anisotropy Probe (WMAP) data. As for the 70 percent of dark energy, I simply went with the cosmological constant or vacuum energy interpretation of dark energy.

But by the end of 2006, I was feeling dissatisfied with the progress of MOG in cosmology, and particularly with my interpretation of the phion field as a Bose-Einstein condensate. I was admitting "dark matter" into my theory, for one may well ask, "What is this phion particle that has not been observed?" One can adopt the point of view that any predicted particle in particle physics and cosmology that has not been detected so far is in fact dark matter. Such particles would include the graviton, the Higgs particle, my phion particle, and of course the numerous supersymmetric partners predicted by standard supersymmetric models of particle physics. However, there is one caveat, that the particle that is identified with dark matter has to be a stable particle, and this excludes many possible candidates for dark matter in the standard model. The phion is a stable particle, like the proton.

There is an important difference between MOG with its phion particle and the standard model of Einstein gravity with dark matter. MOG provides a *gravity theory* with powerful predictive abilities, which is missing in the standard model. MOG explains the astrophysical and cosmological data using the undetected graviton and the phion boson within a modified gravitational theory. In contrast, there is no theory of dark matter, so to speak. Just using Einstein gravity and plugging in dark matter with many adjustable parameters to fit the data does not lead to a true understanding of the nature of dark matter or gravity.

And yet, recently a group of Italian physicists has claimed that they have detected a new light boson from cosmological gamma-ray propagation. Observations using the Imaging Atmospheric Cherenkov Telescopes (IACT) indicate a large transparency of the universe to gamma rays, which was unexpected. This transparency is produced by an oscillation mechanism involving a photon and a new boson with spin 1 and a mass much less than 10^{-10} electron volts. This could in fact be MOG's phion.

My original idea, before getting sidetracked by the Bose-Einstein condensate, had been to remove dark matter entirely from the picture of the universe and just allow visible matter such as baryons in MOG. Whereas the phion field constituted an important part of the theory as a fifth force, I was unhappy with it playing a dominant role in explaining the dark matter in the universe. So I went back to asking the question: Can I develop a form of MOG that amplifies the role of visible baryons, and explains all of the cosmological data available? In this scenario, the phion field would be considered a part of gravity rather than matter. In modified gravity theories, the distinction between an extra field being gravity or matter is blurred.

SUPPOSING A STRONGER GRAVITY

I returned to a possible explanation that I had discovered in late 2004 and January 2005, in which the variation of Newton's gravitational constant G in space and time plays a fundamental role in MOG. The amount of baryons in the hot plasma of the early universe was always multiplied by Newton's gravitational constant in order to get the Friedmann equations within the standard model. Therefore if I amplified the gravitational coupling G in the opaque plasma period before recombination, then I would arrive at the magic number of approximately 30 percent of matter needed to fit the data. In other words, a stronger gravity was taking the place of dark matter.

This was not an arbitrary choice of numbers. In MOG there are equations that determine the variation of G. These solutions for the varying G had to obey the constraint that the gravitational constant be identical with Newton's constant at the time of nuclear synthesis, about one to 100 seconds after $t = 0$. This constraint was needed to fit the observational data on the present abundances of hydrogen, helium, deuterium, tritium, and lithium in the universe. In addition, the variation of G with time had to agree with the upper bound on such a variation that had been set by spacecraft missions such as the Cassini mission in the solar system.

In MOG we retain the experimental fact that the baryons make up only 4 percent of the matter in the universe, and in order to compensate for the lack of dark matter in MOG, we increase the strength of Newton's gravitational constant G. This makes the pull of gravity stronger, leading to *the effect* of the baryons representing about 30 percent of the matter density needed to fit the cosmic microwave background radiation (CMB) data. In gravitational effects in cosmology, we do not measure the density of matter and G separately. We measure the product of G and the density of matter. In nongravitational effects, such as determining the speed of sound in the baryon-photon fluid, the gravitational constant does not enter into the equations. Therefore, observationally we can either say that Newton's constant times the density of matter with the addition of dark matter gives the 30 percent of effective matter needed to fit the cosmological data, or, equivalently, we can say that Newton's constant increased by a factor of seven times the 4 percent of baryons also fits the cosmological data without dark matter.[1]

FINISHING MOG COSMOLOGY

Recently Viktor Toth and I turned our attention to applying MOG to cosmology. We focused on the fundamental fact that the preferred model of cosmology today, the Lambda CDM model, provides an excellent fit to cosmological data but at a huge cost, because about 96 percent of the matter in the universe is either invisible or undetectable or possibly both. Viktor and I felt that this was a strong reason for seeking an explanation of the cosmological observations without resorting to dark matter or to Einstein's cosmological constant. However, the bar is set increasingly higher by recent cosmological discoveries that are neatly explained by Lambda CDM, making it far more difficult to design gravitational theories that challenge the standard model.

As we have seen, alternative gravity theories have to explain successfully the rotational curves of stars in galaxies, the gravitational lensing of galaxies and galaxy clusters, and notably, the angular acoustic power spectrum of the CMB, the matter power spectrum of the CMB and galaxy surveys, and the recent observation of the acceleration of the universe. Armed with the new predictive power of MOG, we now know that it can fit remarkably well all the observational data from the Earth through globular clusters within our galaxy, galaxy rotation curves, clusters of galaxies, and the colliding bullet cluster. Viktor and I also discovered that MOG can explain all the present cosmological observations that go beyond the astrophysical data. That is, we were able to demonstrate that MOG produces an acoustical power spectrum and a matter power spectrum in agreement with the WMAP and balloon data and galaxy surveys. MOG also produces a luminosity distance relationship for supernovae that is in good agreement with the supernovae observations. All of this was accomplished without dark matter and with an explanation of the supernovae data and dark energy arising from the MOG fields, without explicitly including Einstein's cosmological constant.

The key difference between MOG and the standard model of cosmology is that MOG has a varying gravitational constant that depends on distance and time. In developing the MOG cosmology, Viktor and I applied the field equations of MOG to weak gravitational fields and based the cosmological predictions on an exact numerical solution of the MOG field equations.[2]

A COSMOLOGICAL TEST DISTINGUISHING MOG FROM DARK MATTER

When Viktor and I calculated the matter power spectrum and the growth of structure in the MOG cosmology, we made the amazing discovery that in a universe where matter consists only of baryons (and some detectable neutrinos), the curve in the matter power spectrum shows large oscillations. These predicted oscillations are unavoidable in such a universe. In contrast, in a universe dominated by cold dark matter, such oscillations do not occur in the matter power spectrum. When we tried to fit the baryon-dominated universe to the matter power spectrum data without MOG and using just Einstein gravity, the predicted oscillating baryon matter power spectrum did not fit the data at all. When we used the MOG prediction, with just baryon matter and stronger gravity, the oscillating power spectrum fitted very neatly into the SDSS galaxy data, giving a remarkably good fit within the error bars. When we fitted the cold dark matter power spectrum prediction, it was the standard model's smooth curve going through the SDSS data, which contrasted to the wiggly scallops of MOG's fit to the data.

Why does MOG predict an oscillating pattern in the galaxy fluctuations while the standard model does not? Baryons interact with each other, while dark matter does not interact either with other supposed dark matter particles or with baryons. Baryons are like bouncing balls. They hit each other, and when they fall to the ground, they bounce. Dark matter, however, is more like a ball thrown down a bottomless well. It hits nothing, and therefore causes no oscillations. Since in the standard model, dark matter is far more abundant than baryons, it overwhelms and smothers the oscillations of the baryons. In MOG, however, the baryons are the dominant form of visible matter and

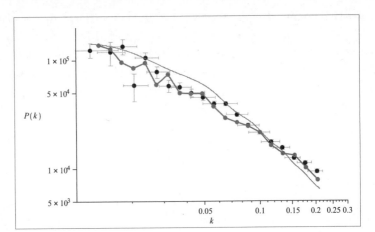

FIGURE 19: The matter power spectrum plotted versus wave number k. The **MOG** prediction is the oscillating curve, while the smooth curve is the Lambda **CDM** prediction. The actual **SDSS** data are shown as dots. When more data points are available, either **MOG** or cold dark matter will be verified. Source: J.W. Moffat and V.T. Toth.

energy, so when they interact, there are plenty of measurable oscillations.[3]

Viktor and I now suddenly realized that we had discovered an amazing test that could distinguish MOG from the standard dark matter cosmology. Within the next two or three years, we expect that the ongoing, very large-scale galaxy surveys such as the SDSS and 2dF surveys will have counted about 100 times more galaxies than at present. This will produce data that are of much finer resolution than today's data, and we will be able to see whether the oscillating MOG baryonic power spectrum is the true description of the data, or whether it is the smooth curve predicted by the Lambda CDM model.

During discussions with physicists about MOG and dark matter, the question inevitably arises: Is it possible to find any way of distinguishing modified gravity theories and cold dark matter observationally? This is ultimately how physics works. A physical theory predicts a result that becomes a test, and an experiment decides between competing theories. Viktor and I had made the startling discovery that indeed such a test was at hand in the near future. Which theory will win? Who will break out the champagne?

Of course, we hope that MOG wins. However, if cold dark matter wins, we will still have achieved something significant: We would finally arrive observationally at a convincing reason as to why dark matter exists and why we should keep Einstein gravity. On the other hand, if MOG wins, this extraordinary result would change the future of astrophysics and cosmology, by telling us that Einstein gravity must be modified to fit the data.

I anticipate, however, that if the MOG prediction wins out, then some dark matter theorists will publish an explanation for the dark matter model that *does* produce oscillations in the data, as this is what has happened in the past throughout the story of dark matter. Whenever some strange and unacceptable phenomenon popped up in the astrophysical and cosmological data that did not agree with what was expected from dark matter models, some theorist would cook up a contrived explanation—like further refinements to Ptolemy's epicycles—to try to get around the problem.

COMPUTER MODELING OF GALAXY GROWTH

Large-scale surveys of galaxies and clusters of galaxies tell us how they are distributed across the sky. Large-scale computer calculations called "N-body simulations" allow thousands of simulated particles to evolve from the early universe of the CMB fluctuations into the late-time universe, to see whether the end result predicted by theory matches the reality of present-day large-scale surveys of galaxies. Extensive studies of this kind have revealed that when only visible baryon particles are used in the N-body simulations, that is, without dark matter or MOG's stronger gravity, the simulation does not agree at all with the observed large-scale structure of the universe. Including neutrinos as dark matter does not help the situation.

There are two theories of galaxy formation. One is the bottom-up model that assumes that the primordial seeds in the CMB clump due to the pull of gravity and form protogalaxy structures about the size of Earth. Eventually these protogalaxies form normal-sized galaxies and clusters of galaxies. A competing model was proposed by the celebrated Russian cosmologist Yakov Zeldovich, who suggested a top-down model starting with large pancake-shaped structures containing galaxies, which due to the pull of gravity spun out into smaller galaxy-sized structures that eventually became the galaxies we see today. But since this scenario for galaxy formation fails to fit the data, it has fallen into disfavor. When undetected dark matter particles such as heavy WIMPs (weakly interacting massive particles) are used in N-body simulations for the bottom-up model of galaxy and cluster formation, then the simulations produce a result that is remarkably close to what is seen in the sky survey data.

It remains to be seen whether computer model N-body simulations based on MOG and containing only baryons and photons can fit the data from large-scale surveys of galaxies. Hopefully these computer models will help to verify MOG by producing realistic descriptions of the large-scale structure of the universe as observed in galaxy surveys without exotic dark matter.

Chapter 14

DO BLACK HOLES EXIST
IN NATURE?

E instein was not happy with the black hole solutions of his field equations. He considered them unphysical because of the strange behavior experienced by observers near a black hole event horizon, and especially because of the ugly singularity at the heart of the black hole. In fact, Einstein abhorred the singularities in his gravity theory—both the singularity that occurs at the beginning of the universe in the big bang model and the singularity within black holes. He hoped that if he succeeded in his quest to discover the true unified theory, then the singularities in his gravity theory could be eliminated.

In his autobiographical notes for his seventieth birthday, Einstein asked the question: "What are the everywhere regular solutions of these equations?"[1] A literal translation of Einstein's original German conveys his meaning better than the common English translation "regular solutions." Einstein was searching for *singularitätsfreie Lösungen*, literally "singularity-free solutions" to his unified theory equations.

EINSTEIN GRAVITY WITHOUT BLACK HOLES

Recently I have been able to discover an exact, nonsingular solution of Einstein's field equations in general relativity—that is, a solution that does not lead to *any singularities*—provided that one assumes there is a permanent exotic field energy permeating all of spacetime. The solution does not possess a black hole event horizon. Of course, by postulating the existence of such exotic field energy throughout spacetime, we have introduced a kind of modified Einstein gravity. However, it is of a simpler form than that of my

Scalar-Tensor-Vector Gravity (STVG) in MOG, which solves astrophysical and cosmological problems without dark matter and explains the origin of dark energy. In any case, this remarkable nonsingular solution for general relativity may herald a new understanding of Einstein's theory and its predictions for collapsing stars. Perhaps Einstein would have been pleased to learn that his theory need not lead to unphysical singularities.

A field such as a scalar or vector field in Einstein's gravity theory that has a negative energy density or pressure near the center of a collapsing star can prevent the star from collapsing to a singular infinitely dense point; it will have a singularity-free exterior gravitational field or spacetime geometry without a Schwarzschild event horizon. For large enough distances from this grey star, the solution reduces to the Schwarzschild solution and therefore agrees with the current gravitational experiments. But when we revert to Einstein's gravity theory without an exotic field energy permeating spacetime, then the solution of Einstein's vacuum equations yields the famous Schwarzschild solution of 1916, which leads inevitably to the existence of black holes.

WHAT IS THIS EXOTIC ENERGY?

The new concept of dark energy or vacuum energy—which was needed to explain the discovery of the accelerating expansion of the universe—has changed our understanding of the nature of matter and energy. Physicists have not yet even agreed on what to call this new energy. It is variously referred to as the vacuum energy, dark energy, or Einstein's cosmological constant. I think of it as a new "field energy" that has negative pressure and density. This exotic field energy permeates all of spacetime, in the interior of stars as well as in empty space. It can be thought of as a new kind of ether that exists universally in spacetime, and it does not violate the symmetries in gravity such as local Lorentz invariance or general covariance.

This new field energy can exert a strong repulsive force as a body collapses and it can overcome the attractive force of gravity, however large the mass of the initial body before collapse. This energy, which has not been entertained to exist in the standard model of chemical and nuclear composition of stars, can change the whole concept of whether black holes actually do form under the collapse of astrophysical objects. Indeed, the negative pressure of this new kind of energy will violate one of the assumptions underlying the Hawking-Penrose theorem that demands that a collapsing body will form an event horizon and an essential singularity at its center.

ALTERNATIVES TO BLACK HOLES

Other physicists have contemplated changing the black hole paradigm. In 2000, the American Nobel laureate Robert Laughlin in collaboration with George Chapline and David Santiago proposed a model of a dark star that had an exterior Schwarzschild solution, but a condensed matter phase transition on a thin shell of matter on the surface of the dark star prevented the black hole event horizon from forming.

In 2004 in a paper published in the *Proceedings of the National Academy of Sciences,* Emile Mottola and his collaborator Pawel Mazur described a "gravastar" (gravitational vacuum energy star) or "dark star." This purported object has a strong vacuum energy with negative pressure preventing its interior from collapsing to a black hole singularity, and it does not have a black hole event horizon. This is because the gravastar is joined at its surface to the Schwarzschild solution by a thin shell of matter, and a similar condensed matter phase transition as Laughlin's occurs at its surface. When Mottola and Mazur first submitted their paper to *Physical Review D,* the reviewers scorned the ideas and rejected the paper, claiming that such a gravastar could not form in Einstein's gravity theory.

Physicists who champion quantum gravity commonly believe that we cannot describe the center of a collapsed star with a classical theory of gravity. This is where quantum mechanics comes into play: A quantum theory of gravity becomes necessary to describe what happens to matter at very short distances. Indeed, there are several versions of quantum gravity and string theory that claim to avoid the unphysical infinity of the density of matter at the center of the black hole while retaining the event horizon. The validity of such solutions is controversial, however, for in order to prove the absence of singularities, we need a universally accepted solution to quantum gravity, which is not available today.

MOG'S GREY STAR

What if the dense object at the center of the galaxy did not have an event horizon, but instead was a very massive object containing billions of stars, forced by mutual gravitation into a small, dense, stable object? Then this dense object would have a surface much like the surface of a star, and stars would not disappear into it through an event horizon. If such a supermassive object were stable, with a mass of about one million solar masses, then we would have a different, alternative picture of what those very massive, dense objects at the centers of galaxies might be. Due to the strong gravitational pull of the dense conglomeration of stars, light would have trouble escaping, and it would be difficult to see the dense object with optical telescopes. But enough light would escape to make the object dark grey, rather than black.

This is in fact the super-grey object that solutions of the MOG field equations could predict. It would sit at the center of the galaxy, comparable to the giant black hole predicted by Einstein's gravity theory. When we solve the field equations of MOG, and include the massive phion field coupled to gravity and matter, and the variation of G, then the static, spherically symmetric solution of the field equations exterior to an astrophysical body may well not be the same as the Schwarzschild solution in general relativity. I call such a possible MOG solution—the massive object at the centers of galaxies—a "super-grey star."

A similar solution for a collapsed dark companion in a binary system could also be a "grey star" and not a true black hole as predicted by Einstein's gravity theory. This potential grey star solution in MOG describes an object with a surface like an ordinary star that can be about the same size as a black hole in Einstein's gravity theory. It is held together by a stabilizing repulsive force produced by the exotic field energy, which prevents its collapse to a black hole.

MOG's varying gravitational coupling constant G, in addition to the new repulsive field energy and MOG's phion field, makes it possible to obtain a solution of the generalized gravitational field equations for a static, spherically symmetric field around a massive body like a collapsing star *that is finite, nonsingular and does not have an event horizon.* Thus in MOG solutions may exist for which there is no event horizon surrounding the final state of a collapsed star, and there is no essential singularity at its center.

Let us begin again with a star whose nuclear fuel has been depleted, and that is massive enough so that the force of gravity makes it collapse. During the collapse of the star, the dark field energy within the star begins to produce enough negative pressure to halt the collapse. This negative pressure eventually reaches a balance with gravity, preventing the collapsing star from forming a singularity at the center with infinite density. In other words, the object we have in MOG is not a black hole but a dead grey star with a surface like that of a normal dead star. The gravitational field exterior to this dense object is not the event horizon of the Schwarzschild solution of general relativity, but another solution to MOG, which does not have an event horizon. The object is not black, like a black hole, because some visible radiation can escape. But the amount can be tiny, so that from a great distance it would be difficult to detect it. Nevertheless, because some information can escape, the whole sticky issue of information loss disappears! Nor is the object a hole in spacetime, as a black hole is envisaged, but rather it is a dark, massive, dense—but not infinitely dense—stellar object.

A large part of the relativity community is in denial—refusing even to contemplate the idea that black holes may not exist in nature, or seriously consider the idea that any kind of new matter such as the new putative dark energy can play a fundamental role in gravity theory. The strange behavior of black holes, such as the reversal of time and space as you enter the black hole through the event horizon, is viewed as just another "relativity phenomenon" like the twin paradox or the slowing down of clocks, and is an accepted part of the counterintuitive world of relativity.

However, as the MOG grey star solution shows, the universe may not be as weird as black hole physics has led us to believe for decades. You do not need to postulate strange nonlocal hologram behavior or black hole complementarity principles, or invoke quantum entanglement to describe the normal physics of the final state of a star's collapse. What is needed, though, is the same exotic field energy or dark energy that is invoked to explain the accelerating expansion of the universe. Moreover, quantum physics can also lead to negative field energy and negative pressure as a collapsing star approaches its center, thereby preventing a singularity from forming.

At the time of this writing, recent theoretical research such as my work on MOG shows that it is possible that nonsingular collapsed stars exist, but these solutions must still be investigated further. There are also fundamental unsettled issues of both a mathematical and physical nature that prevent us from fully excluding black holes and their strange properties. But the subject is, in my opinion, not a closed one. As I explained in Chapter 5, all the evidence so far for the existence of black holes is circumstantial. When one truly believes in the existence of black holes, as most astrophysicists do, then that circumstantial evidence probably appears very compelling. But I believe that further research, however difficult to perform, should be carried out to reveal the truth.

Chapter 15

DARK ENERGY AND THE ACCELERATING UNIVERSE

During 2007, the U.S. government provided significant financial support to a U.S. scientific task force to investigate the existence of dark energy and its consequences. This represents a major part of the future funding for space science research. It underscores how important the scientific community considers the quest to understand dark energy since the dramatic supernovae results in 1998 suggesting that the expansion of the universe is accelerating.

Indeed, since then, there has been an avalanche of scientific papers attempting to explain this startling and unexpected phenomenon, both by theorists and experimentalists. Many of the theoretical explanations ascribe the apparent acceleration of the universe to dark energy—a mysterious new kind of energy permeating the universe. "Dark energy" is somewhat of a misnomer because it does not immediately distinguish itself from dark matter, which is also of course a form of dark energy, since matter and energy are equivalent. Still, any theoretical attempt to discover a cosmological model must confront the supernovae data and the mystery of the dark energy.

As we have seen, MOG has been able to fit astrophysical and cosmological data successfully over fourteen orders of magnitude in distance scales. But this is still not enough to convince the physics community that MOG is the correct replacement of Einstein's general relativity. There remain the all-important supernovae data and the question arising from them: Why is the universe accelerating—if it is!—and what exactly is causing the acceleration? Why is the acceleration occurring in our epoch, and does this suggest an anti-Copernican universe?

IS THE UNIVERSE REALLY ACCELERATING?

During 2004 and 2005, when I was investigating the accelerating expansion of the universe, I published papers attempting to explain it based on exact inhomogeneous solutions of Einstein's gravitational field equations without any modification of his theory. Unlike the techniques of perturbation calculations, an *exact solution* of field equations in physics does not ignore any contributions that are difficult to calculate. The standard FLRW equations, which are exact solutions, assume that spacetime is homogeneous and isotropic, whereas in an inhomogeneous solution we drop this assumption.

My motivation for this work was simple: If general relativity could explain the phenomenon of the accelerating universe, why look elsewhere? It appeared that I might possibly succeed in fitting the data to Einstein's theory, and manage to avoid the controversial "coincidence" problem, which asks the loaded question: "Why is the acceleration of the expansion of the universe happening *now*, when we can observe it?"

Others have also attempted to use general relativity to explain the acceleration of the universe. In a paper in *Physical Review Letters* in 2005, Edward ("Rocky") Kolb and his Italian collaborators proposed a scenario in which there is, bluntly speaking, no dark energy at all, and no need for Einstein's cosmological constant or vacuum energy. The idea is that small perturbations or modifications of the exact homogeneous and isotropic background spacetime of the FLRW universe—that is, *approximate* solutions of the equations—would naturally introduce inhomogeneous contributions that would cause the expansion of the universe to apparently accelerate. A press release by the American Physical Society accompanied the publication of the paper, claiming that the authors had explained the accelerating expansion of the universe.

Papers criticizing this approach followed promptly in the electronic archives, at the rate of at least one every other day for about two weeks. The problem was the perturbative methods that Kolb and his collaborators had used. That is, they assumed that the effects obtained from an inhomogeneous distortion of spacetime were small. This led to all sorts of technical problems. Moreover a famous equation in cosmology published by the Indian cosmologist Raychaudhuri in the 1950s, and later interpreted by Stephen Hawking and George Ellis, says that the universe cannot locally show an acceleration of expansion if its matter density is positive. This would seem to contradict the claims of Kolb and his collaborators, who needed to retain positive matter density.

By solving Einstein's field equations *exactly*, we can get a more precise answer to the question of why the expansion of the universe appears to be accelerating than one can accomplish with perturbation theory. My papers, among others, investigated this idea. One of the original cosmologists to suggest using exact solutions of Einstein's field equations to explain the accelerating universe was Marie-Noëlle Célérier at the Meudon Observatory near Paris. In 2000 she published a paper proposing that late-time acceleration of the universe could simply be reinterpreted as late-time inhomogeneous structure: When an observer looks into a cosmic void with galaxies surrounding it, the normal expansion of the spacetime inside the void is faster than outside the void, so the observer has the *impression* that the universe within his horizon is accelerating.* If the galaxies and voids with their relative motions were removed, then the observer would see—according to Célérier, myself (using only Einstein gravity), and others—that the universe was actually *decelerating*. Célérier claimed that the dimming of the light from supernovae, the original signal of the accelerating expansion, was caused by the light passing through the inhomogeneous medium of galaxies, clusters, and voids.

This is the only sure way I know to counter the anti-Copernican idea that the universe is accelerating in our epoch—to get rid of the problem entirely! Of course, we can never remove the large-scale structure of late-time galaxies, clusters, and voids, so we will never be able to prove by observations that this interpretation of the supernovae data is true. This, then, opens the door to the argument that we *do* live in a special time and place in the universe—one in which the universe is accelerating its expansion—which is counter to the Copernican and cosmological principles.

What is at stake here is that the cosmology community generally assumes that the density of matter is homogeneous throughout spacetime. Einstein put forth this idea in his original paper on cosmology in 1917, and it is known as the cosmological principle. However, we know from observations and large-scale surveys of galaxies that the universe has not been uniform and homogeneous in our late-time epoch during the last nine billion years. Indeed, instead of uniformity we see large distributions of galaxies and large voids.

In my attempts to fit the supernovae data to Einstein's theory, I sought to understand this observational inhomogeneity, which should have a definite impact on the rate of the universe's expansion. I studied an exact solution of Einstein's field equations found originally by Lemaître and investigated further by Richard Tolman and Hermann Bondi. This solution, called LTB

---

* Telescopes have observed giant voids larger than 100 million parsecs.

for its originators, assumes that the universe is spherically symmetric and has a center, much like the spherical symmetry of a blown-up balloon.* Since the LTB solution is exact, we can use it to obtain precise predictions about the expansion of the universe. It turns out that in the LTB solution, additional contributions occur due to the inhomogeneous distribution of matter at late times that can cause the expansion of the universe to speed up with a positive matter density and pressure, provided that we perform a spatial averaging of the inhomogeneous matter distribution. However, the mathematical story is not simple.[1]

As early as 1991–1992 and again in 1995, I published extensive investigations of the LTB model and its implications for the large-scale voids in cosmology—all preceding the supernovae data. This work was published in the *Astrophysical Journal* with my student Darius Tatarski. We discovered that the existence of voids in the LTB model of the universe had important implications for cosmology, anticipating the possibility of the dimming of the supernovae light and the interpretation that the universe is accelerating.

Many physicists have criticized the use of a spherically symmetric solution such as the LTB solution because it assumes a center to the universe. However, I recently studied a more general inhomogenous solution to Einstein's field equations based on a solution published in 1977 by Duane Szafron, then a postdoctoral fellow in the mathematics department at the University of Waterloo. This solution is not spherically symmetric—it has no center to the universe—yet it can still explain the acceleration of the expansion of the universe.

In ideas like mine, Célérier's, and many other cosmologists', the unexpected dimming of distant supernovae is explained by the idea that intervening galaxies, clusters of galaxies, and voids form a medium like air or water through which light travels. In other words, in these models the universe is not accelerating at all; the light from distant supernovae just has a hard time getting to us, so it *appears* that the supernovae are farther away than they actually are. The idea of using the late-time inhomogeneity of the universe to avoid the problem of the accelerating universe and dark energy entirely is elegant. But it leaves many imponderables. It is still a controversial issue whether these models can continue to fit the supernovae data as more accurate data become available; and there exists the competing standard Lambda CDM model, and as we shall see, an explanation following from MOG.

---

* The authors did not specify where the center of the universe was. This feature of the model is actually an artifact due to the assumption of spherical symmetry.

AN ACCELERATING UNIVERSE WITH DARK ENERGY OR ALTERNATIVE MODELS

Many cosmologists accept the idea that the universe is indeed accelerating, and they explain it using Einstein's cosmological constant or vacuum energy, or the mysterious dark energy. In this explanation, the density of energy remains positive but the pressure is negative. For ordinary matter, both the pressure and the density are positive. But the energy of the vacuum has a unique interpretation in physics. To support the vacuum energy in space-time, its pressure has to be negative, which means that it is "pulling," rather than "pushing." Think of a piston capping a cylinder containing a vacuum. When the piston is pulled farther out, more vacuum is produced. The vacuum within the cylinder then has more energy, which must have been supplied by the force pulling on the piston. Picture now that the vacuum is trying to pull the piston back into the cylinder. To do this, the vacuum must have a negative pressure (relative to outside), since a vacuum with positive pressure would tend to push the piston out. The vacuum energy in space consists of quantum fluctuations, while the region outside the vacuum energy has zero density and pressure. The reader may be wondering what it means to have something "outside" the vacuum energy, because in space there is no "inside" or "outside." Frankly, physicists do not understand this problem either; it is one of the deep reasons that we do not understand the nature of dark energy.

Nevertheless, the standard explanation for the acceleration of the universe involves dark energy or the dark, repulsive force. We recall that this dark energy with negative pressure is believed to be uniformly distributed throughout spacetime and it does not clump through gravitational forces as dark matter supposedly does. The dark energy has also been called "quintessence." (Quintessence is from the Latin *quinta essentia*, which means the fifth essence or element. Aristotle referred to it as an hypothesis, and Anaximander [c. 610 BC–c. 540 BC] wrote: "It is neither water nor any other of the so-called elements, but some different, boundless nature."[2])

We recall that another possible explanation for the acceleration of the universe was published by Georgi Dvali, Gregory Gabadadze, and Massimo Porrati. Their modification of Einstein's equations adds a fifth dimension, which generalizes the Friedmann equations of cosmology so that as the cosmological distance scale increases in size, a repulsive antigravity force appears and drives the universe into its period of acceleration. But again, such a theory has potential serious problems with instability and it may also lead to disagreements with solar system observations. Moreover, the introduction of an extra dimension, as always, makes a model difficult to verify because no one has ever observed an extra dimension.

Yet another explanation for the accelerating universe was published by Sean Carroll, Damien Easson, Mark Trodden, Michael Turner, and others in *Physical Review D* in 2005. They modify Einstein's gravitational theory by adding a function of the curvature of spacetime to the theory's basic action principle. By a suitable choice of adjustable parameters, the generalized Friedmann equations can be made to accelerate the universe without a cosmological constant. There has been a proliferation of papers investigating this kind of alternative gravity theory. A major problem, yet again, is that the models can display the demon instability problems and may not be consistent with solar system observations. Indeed, Michael Seifert of the University of Chicago has shown that the particular model proposed by these authors would only allow the sun to shine for fractions of seconds. Finally, neither this alternative model nor the Dvali-Gabadadze-Porrati model can account for the coincidence problem: Why is the acceleration of the universe happening in our epoch?

MOG AND THE SUPERNOVAE DATA

What about a unified description of the universe based on MOG? A varying gravitational constant and the energy density and negative pressure induced by the phion vector field can lead to a unified description of modified gravity without dark matter, and *also* to an explanation for the dark energy without Einstein's unwanted cosmological constant. It would be more satisfying to explain the acceleration of the universe by means of the field equations of MOG, since MOG already fits other cosmological data that general relativity has failed to fit without dark matter or the cosmological constant. If MOG could explain the supernovae data too, then we would arrive at a unified theory of cosmology.

Viktor Toth and I recently published an application of MOG cosmology in which the MOG field equations actually predicted an accelerating expansion of the universe that conforms to the supernovae data. In this work we assumed that visible baryons constitute the only matter in the universe.

MOG's repulsive force has a different physical origin from Einstein's cosmological constant. The latter is caused by negative-pressure vacuum energy in the universe, while in MOG this vacuum energy is a consequence of solving the MOG field equations that predict a repulsive, antigravity component within gravity itself. In Einstein's gravity theory, based on a homogeneous FLRW universe and without a cosmological constant, there is no such antigravity, so the universe would decelerate in the future. When you put the cosmological constant into Einstein's equations, you can get an accelerating expansion of the universe, but you also get all the problems associated with the cosmological constant, such as the calculation based on standard particle physics that produces a number 10^{120} times larger than is permitted by observation.

In MOG, however, the antigravity component is built into the theory's equations from the beginning—it is an aspect of spacetime or gravity itself—and it does not have the adverse consequences of Einstein's cosmological constant. Its effect is to push the fabric of spacetime to expand faster and faster. This explanation of the accelerating universe fits into the theory's larger picture with all the other consequences for astrophysics, the solar system, and cosmology obtained from the MOG field equations. MOG has a unified explanation for all the presently available observational data, while it removes the need for dark matter. In other words, what we observe in the cosmos is all due to gravity and visible matter. We consider this scenario a successful alternative to the standard Lambda CDM cosmology.

What about the anti-Copernican coincidence problem in MOG? Given the acceleration of the universe in MOG, then the theory should have enough predictive power to tell us that the universe will inevitably be accelerating in the present epoch. At this writing, this is still an unresolved issue in MOG.

DARK ENERGY IS STILL AN OPEN QUESTION

Let me say again that there is no consensus among cosmologists about what is causing the acceleration of the universe, or even whether the acceleration is a real effect. There are now many published explanations for this unexpected phenomenon, including MOG's solution, and they all claim to fit the supernovae data. How do we distinguish among these different explanations and decide which one is correct? A theory has to produce testable predictions that uniquely differentiate it from its competitors. In MOG, we have the theory's successful predictions for astrophysics and cosmology without dark matter, but we do not yet have a *definitive* prediction for the acceleration phenomenon.

Even with future, better observational data for the supernovae, it is far from clear whether we will ever obtain a definitive explanation for the possible acceleration of the universe. This mystery and the elusive dark energy may remain with us for a long time.

Chapter 16

THE ETERNAL UNIVERSE

One of the most disturbing features of the standard big bang model is the singular beginning of the universe at time t = 0, when the density of matter is infinite. Moreover, having a beginning to the universe can imply the existence of a creator. Most cosmologists are not happy bringing creation, religion, and mysticism into their calculations, as such ideas are considered outside the scope of science.

The father of the big bang, Georges Lemaître, was strongly opposed to mixing religion and science. He held this belief in spite of the fact that he was a Catholic priest and a member of the Pontifical Academy at the Vatican. Although Lemaître's expanding universe model began with his primeval atom or cosmic egg, he did not in any way suggest that his theory constituted proof of the creation story in the book of Genesis. Recall that this was in contrast to Pope Pius XII, who endorsed the big bang theory. On November 22, 1951, the Pope said (in so many words) that the explosive beginning to the universe proved the existence of God:

> What was the nature and condition of the first matter of the universe? The answers differ considerably from one another according to the theories on which they are based. Yet there is a certain amount of agreement. It is agreed that the density, pressure and temperature of primitive matter must each have touched prodigious values . . . Clearly and critically, as when [the enlightened mind] examines facts and passes judgment on them, it perceives the work of creative omnipotence and recognizes that its power, set in motion by the mighty *fiat* of the Creating Spirit billions of years ago called into existence with gesture of generous love and spread over the universe matter bursting

with energy. Indeed it would seem that present-day science, with one sweep back across the centuries, has succeeded in bearing witness to the august instant of the *fiat Lux,* when along with matter, there burst forth from nothing a sea of light and radiation, and the elements split and churned and formed into millions of galaxies.[1]

Modern cosmologists view the big bang somewhat differently, and more as a metaphor, than Lemaître's original idea that the universe began with a huge explosion. Today's cosmologists interpret the beginning of the universe at t = 0 as a very dense, hot, and small place, but not necessarily a place where an explosion occurred. However, many cosmologists still accept that a singularity with infinite density did occur at t = 0, as predicted by general relativity. The cause of the beginning of the universe, what came before the beginning, and the expansion of the universe from t = 0 in the standard model are not well understood. Indeed, there is now a minority of cosmologists who question a beginning to the universe at all; instead they favor a cyclic model with a series of expansions and contractions.

This singularity at the heart of the big bang not only implies a creator. With its infinite density and temperature, it also means that the laws of physics break down at the beginning of the universe. If you were to program the evolution of the universe on a computer, the computer would crash when attempting to compute how the universe began with a singularity. Some physicists have argued that somehow a satisfactory quantum gravity theory could avoid the singularity at t = 0 in the standard big bang model, but so far no satisfactory quantum gravity theory has been shown to solve this problem. Even among those cosmologists who might believe that the universe was created by God, the singularity at t = 0 remains a vexing problem.

THE MOG COSMOLOGY

The MOG universe has a beginning but not an end. It continues eternally. There is no explosion of matter from an infinitely dense singularity. The beginning of the MOG universe is a time of vast stillness and emptiness before the universe evolves dynamically. In contrast to the standard big bang model, there is neither matter nor energy at t = 0. Gravity—the curvature of spacetime—is also zero because there is no matter. And the Hubble constant H, which governs the expansion of the universe, is also zero in the primal vacuum. At MOG's t = 0, the universe does not explode or inflate, but instead it stands still.

The MOG universe at t = 0 is a very important and unstable place. There is no singularity in the solution of the MOG field equations at t = 0. In fact, no singularity is ever encountered in the MOG cosmology. In the neighborhood of t = 0, as the universe stands poised, quantum mechanical fluctuations bring particles of matter and energy into being. Their presence forces curvature into the spacetime. Soon MOG's gravity comes into play, the Hubble parameter becomes nonzero, and the universe begins expanding. The density and pressure of matter form the hot, dense plasma of quarks, electrons, and photons, the hot cosmic soup that we recognize from the conventional big bang model. Pope Pius XII's statement (". . . when along with matter, there burst forth from nothing a sea of light and radiation") is a poetic description that actually fits the MOG beginning of the universe better than it does the big bang. For in MOG, matter and radiation came from nothing—the modern physics vacuum. Matter and energy were not already squeezed into an infinitely dense singularity that then exploded.

How does this cosmology come out of the MOG theory? For one thing, the variation of the gravitational strength G in MOG plays a significant role in the early universe, when the geometry of spacetime begins evolving dynamically. This is in contrast to the standard big bang model where there are no extra fields to modify the curvature of spacetime; spacetime in MOG is curved differently than in Einstein's gravity theory.

Secondly, in order to avoid a singularity at t = 0, the MOG cosmology equations known as the generalized Friedmann equations with a zero cosmological constant must satisfy certain conditions. One condition is that there is no matter or radiation at t = 0; neither is there any gravity or curvature of spacetime. The universe is empty, and spacetime is described by the spatially flat Minkowski geometry.

There is some inflation in MOG at t = 0. This is needed mathematically to remove the singular (infinite) behavior of the cosmological solution. This means that the initial expansion from t = 0 is accelerating by a small amount, like the sudden acceleration of a car after stopping at a stop sign. MOG's inflation at t = 0 is not nearly as large and dramatic as in standard inflationary models.

In MOG at t = 0, matter and radiation are created from fluctuations in the vacuum. The temperature of the resulting hot plasma must be very high in order to yield the temperature of 2.7 degrees Kelvin today, after spacetime has been stretched. In contrast to the singular big bang model, these high densities and temperatures need not reach those of the Planck density and temperature where quantum gravity becomes important. Thus we may not need quantum gravity as the operative mechanism to do away with the singularity at t = 0. MOG accomplishes this in four-dimensional spacetime without the need for ill-understood quantum gravity theories or string theory.

ENTROPY AND THERMODYNAMICS

A limiting factor when one considers how—or whether—the universe began is the second law of thermodynamics, which states that entropy, or disorder, always increases in the universe.[2] For example, heat always passes from a hot body to a cool one and not the other way around. A well-known analogy is with a dinner plate falling off a table and crashing to the floor. We are moving from a highly ordered state, namely the dinner plate, to a highly disordered one, the porcelain fragments. The thermodynamical arrow of time points in the direction of increasing entropy and disorder. We cannot reassemble the broken plate back into its initial state on the dinner table. We observe a universe that is asymmetrical in time. That is, there is an arrow of time pointing only to the future.

When we think about the beginning of the universe, we cannot violate the second law of thermodynamics, which is as hallowed a law of physics as the conservation of energy. MOG's very special place at t = 0 is when entropy is at a minimum value, because there is no matter or gravity at all. The entropy only increases from t = 0 as matter and radiation form. So in moving through time, entropy continually increases to the present time and beyond, until the universe will eventually dilute to a totally disordered state that continues forever.

This MOG scenario can be described as an "order-disorder" universe, as opposed to an "order-order" or a "disorder-disorder" universe. By order-order, for example, we mean that the universe begins in an ordered state and with minimal entropy, and evolves toward a similarly ordered state with minimal entropy. The order-order and disorder-disorder scenarios both risk violating the second law of thermodynamics.[3]

An interesting feature of the MOG cosmology is that at t = 0, entropy can just as easily increase toward negative time as positive time. Then, entropy will continually increase into the infinitely distant past or negative time. Thus t = 0 in MOG is a special time when the universe can expand both into positive time and into negative time. Because there is no singularity, the infinite past can be joined to the infinite future; a singularity would cause an obstruction between the past and the future. This is because the Friedmann equations, which apply to MOG as well as to general relativity, are invariant in time. That is, you cannot tell from the equations which way time is running—forward or backward. Two identical universes grow

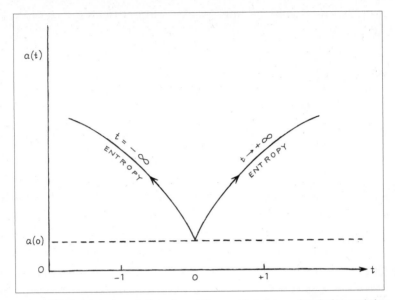

FIGURE 20: **MOG's beginning of the universe has no big bang or singularity, and the second law of thermodynamics is never violated, as there is no going back in time. At the beginning of the MOG universe, both negative and positive time evolve with increasing entropy.**

outward from t = 0 in MOG, one in positive time and one in negative time, and we cannot tell which one we are inhabiting. What these two universes accomplish for us, however, is that they prevent us from violating the second law of thermodynamics.

We picture the eternal universe as two mirror images in time, both representing an asymmetrical arrow of time. In order for the solution of the MOG cosmological field equations to be nonsingular at t = 0, the cosmic scale a(t) must be constant at t = 0. In a cosmological model invoking a beginning to the universe, we have to impose initial and final boundary conditions. The initial condition is that the universe starts in an ordered state with minimal entropy, and the final condition is that it ends in a disordered state. We also impose a condition that prevents us from reversing these two asymmetrical universes. In other words, we cannot approach t = 0 from negative time and come to a "big crunch."

Some physicists, such as Roger Penrose, argue that even if we contemplate, as we do in the MOG cosmology, not having a big crunch, the collapse of matter to black holes and their singularities would still result in a violation of the second law of thermodynamics if we pass through the black hole event horizon. However, MOG, as we recall, can do away with the black hole and its concomitant singularity and event horizon. The removal of the black hole singularity in MOG goes hand in hand with the lack of a singularity at the beginning of the universe. This way the second law of thermodynamics is maintained throughout the evolution of the universe.

There is something very appealing in the idea of a cyclic, eternal universe—one that expands and contracts endlessly, going from big bang to big crunch and back again, forever. This kind of cyclic cosmology was first envisioned by Richard Tolman and others in the 1930s but eventually went out of fashion. Tolman concluded that a cyclic or oscillating cosmology, based on a spatially closed solution, would survive a big crunch and a singularity, only to grow ever bigger and bigger for each new cycle, violating the second law of thermodynamics at every big crunch. This runaway series of cycles would eventually require an absolute beginning and an end to the universe.

More recently, Paul Steinhardt, Neil Turok, and my colleague at the Perimeter Institute, Justin Khoury, have brought back the cyclic cosmology in their ekpyrotic model.* They envisage a string-brane cosmology in which two branes collide, creating a "fiery" universe, after which the branes part and then eventually collide again, in an infinite number of cyclic collisions. The cosmology does not have an inflationary epoch. Steinhardt and Turok have described this cosmology in a recent popular book, *Endless Universe.*

However, the major problem with the idea of a cyclic universe is that entropy *decreases* as the cycle turns toward the crunch and a new expansion. The MOG equations do allow a cyclic cosmology. I experimented with this scenario, but had to discard it, somewhat reluctantly, because of the serious problem of violating the second law of thermodynamics.

SOLVING THE INITIAL VALUE PROBLEMS

We recall that the big bang scenario has several initial value problems that inflation and VSL (variable speed of light) solved, such as the horizon and flatness problems. How are these resolved in the MOG cosmology?

The horizon problem arises because although the cosmic microwave background (CMB) temperature is uniform across the sky, there are several hundred horizon distances between opposite parts of the cosmic sky. Observers can only communicate with one another as far as the end of their own horizon. Thus in the big bang model, there never would have been a way of causally communicating with a finite speed of light the uniformity of the temperature of the CMB. Unless one postulates a cosmology based on a much faster speed of light in the early universe, the only other explanation

---

* Ekpyrotic is a Greek word meaning "out of fire."

has been that the universe went through an inflationary epoch near time t = 0. MOG has now introduced a third possible resolution of the horizon problem: The very early universe had an infinite horizon—in effect it had *no horizon*—which means that all parts of it could easily communicate with the rest. This is preferable to the inflationary and VSL models because it avoids the breakdown of physics at a singularity at t = 0, which has been a source of criticism of the inflationary and VSL models and the big bang model in general.*

MOG also resolves the flatness problem. The WMAP (Wilkinson Microwave Anisotropy Probe) data show that the universe is spatially flat. How can this be, after it has been expanding for almost fourteen billion years? We might expect it to be spatially closed like a sphere or open like a saddle, which are the two alternative possible solutions of the spacetime geometry. In the standard big bang model without inflation, the spatially flat solution would lead, as we recall, to a serious fine-tuning problem. The flatness problem in the MOG cosmology does not even arise because the curvature is zero at t = 0 and remains zero as the universe expands until the present epoch, thus agreeing with the WMAP observations.

MOG AND THE DATA

Recall that Hubble's swiftly receding galaxies "proved" the truth of the big bang model originally. How does MOG explain these data? In MOG, the galaxies and stars are formed several billion years after t = 0, in the matter-dominated phase of the universe, just as in the standard big bang scenario. Since the Hubble constant H is now not zero in MOG, we see recession of the galaxies just as in the big bang model as the universe expands toward the present day.

The seeds of galaxies seen in the CMB are explained in the MOG cosmology without standard inflation. The vacuum fluctuations in MOG's varying G and phion fields in the neighborhood of t = 0 generate the scale-invariant primordial power spectrum that is required to explain the WMAP power spectrum data discussed in Chapters 7 and 13.

How can we interpret the CMB, the fossilized "afterglow of the big bang," in MOG if there *was* no big bang singular beginning? This is only

---

* We must not be confused by the different ways of using the term "infinity." The infinite density associated with the standard model's singularity at t = 0 is an unphysical feature of the cosmology, whereas the infinite time or the infinite horizon in MOG do not lead to a breakdown in physics.

a semantic difference, for MOG and the standard model become the same at the time of recombination—except that MOG does not have dark matter, but a stronger gravity than the standard model. We can merely say that the CMB is the fossilized evidence of the decoupling of matter and radiation, a step in the evolution of the universe. The CMB does not need to arise from a big bang at all. In MOG soon after t = 0, the universe consisted of a very hot radiation plasma of electrons, quarks, and photons. As the universe expanded after t = 0, the radiation plasma cooled down, we have recombination and the decoupling of matter and photons almost 400,000 years after t = 0, the formation of hydrogen atoms, etc., just as in the big bang model. Unfortunately, though, because the opaque curtain of the surface of last scattering obliterates any events before 400,000 years after t = 0, we cannot yet "see" either the standard model's dramatic singular beginning with a huge inflationary increase or MOG's nonsingular, smoother, calmer creation of matter and expansion around t = 0.

Will it ever be possible to tell from observations which version of the beginning of the universe is correct? We need a form of radiation that can lead to direct information about the state of the universe at t = 0. We recall that gravitational waves, unlike photon radiation, interact only weakly with the proton and electron plasma. So gravitational waves *could in fact* give us information about the universe before recombination because they can penetrate the surface of last scattering for the most part intact. It is important to try to detect gravitational waves with projects like LIGO (Laser Interferometer Gravitational Wave Observatory) and LISA (Laser Interferometer Space Antenna). The MOG universe model without standard inflation, the ekpyrotic cyclic model, and the standard big bang model with inflation all have *different predictions* for the intensity and amplitude of the gravitational waves that come from t = 0 and penetrate the CMB curtain. Thus a successful detection of gravity waves could decide among these models.

A less direct way of learning about conditions in the very early universe is to measure the polarization of the CMB radiation when it passes through fluctuations of the gravitational waves. This signal is different from the one caused by radiation passing through the matter and temperature fluctuations in the CMB. If there were no fluctuations or seeds of galaxies observed in the CMB, then all the different potential polarizations of the electromagnetic waves would cancel out due to the uniformity of spacetime. Electromagnetic waves can be vertically polarized or horizontally polarized, corresponding to the up-and-down or back-and-forth "wiggles" of the electromagnetic waves as they move forward in space. When the waves pass through the matter or thermal fluctuations in the CMB, they produce what is called an "E-mode

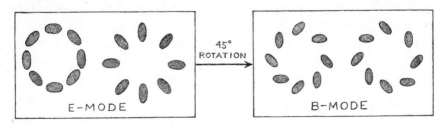

FIGURE 21: The distinctive E-mode and B-mode patterns of polarization could verify one of the several models of the beginning of the universe through the B-mode gravitational signal.

signal," named after the electrically charged particles producing an electric field. This simple pattern is straight lines originating from a central point. On the other hand, the electromagnetic waves passing through the fluctuations of the gravitational radiation cause a complex pattern of polarization, with twisted, circular, vertical and horizontal polarizations all mixed together, creating what is called a "B-mode polarization." This latter polarization is similar to the electromagnetic force fields caused by magnetism. The letter B in this case is a common notation by physicists for the magnetic field.

The E-polarization of the CMB was discovered in the DASI (Degree Angular Scale Interferometry) satellite observations in 2000, which was a further important verification of the existence of the fluctuations in the CMB.* So far, astronomers have not been able to detect the B-mode of polarization in the CMB data. Future NASA satellite missions have been designed to detect the B-mode polarization, particularly the Planck satellite mission to be launched in 2008.

It is amazing to consider that today we might be able, through these gravitational wave experiments, to say what happened in the universe at t = 0. This is an extraordinary time in the history of science, in that we cannot only theorize about the beginning of the universe, but actually study the celestial fossils of how it happened. Just since the 1920s, we have progressed from not knowing whether our Milky Way galaxy was the whole universe to performing observations in space that will tell us what happened fourteen billion years ago.

---

* DASI is a University of Chicago–Caltech project operating from the South Pole.

Epilogue

A s this book goes to press, the community of cosmologists and physicists who believe in dark matter continues to staunchly oppose possible modifications of Einstein and Newtonian gravity. This is despite the fact that dark matter continues to be undetected. At a February 2008 meeting on dark matter and dark energy at UCLA, and at the American Physical Society meeting in St. Louis two months later, experimentalists reported that the search for WIMPs in underground laboratories has come up empty-handed: No dark matter has been discovered. Experimentalists should continue the search for dark matter, however, because it is very important to verify whether it exists. We await the results from the Large Hadron Collider at CERN, which begins operation this year.

Perhaps I have to some extent adopted Einstein's stance in 1905, when he was developing special relativity. He mentioned in passing in his celebrated paper on special relativity that efforts to detect the ether had failed. He then ignored the possible existence of an ether and developed his theory without it. Similarly, I have adopted the point of view that dark matter does not exist, and I have asked the question: Can we develop a new theory of gravity that can successfully explain the extensive observational data through gravity alone?

I do not expect that the present physics and cosmology community will give up its belief in the existence of dark matter easily, even though dark matter may never be detected. Giving up Einstein's theory of gravity is simply unacceptable to many in the community. It may take a new generation of physicists to view the evidence with unclouded eyes.

The attempt to draw back the dark veil that hides the true workings of nature has been a passionate, thrilling experience for me during the past several decades. Developing modified gravity with my collaborators has been one of the most exciting experiences in my physics career, and MOG will continue to be of central importance in my future research. In spite of the blind alleys we have encountered during the quest—the several failed attempts to understand MOG, which I have recounted in this book—we continue the patient pursuit of discovering and fleshing out this theory of gravity that succeeds well in explaining nature.

I hope that this book has succeeded in describing to you, the reader, how difficult it is to try to wrest precious, fundamental secrets from nature. That quest can be compared to climbing a mountain, and when reaching the peak, seeing another higher mountain that tempts us to ascend to even greater heights. And when we do reach the higher peak, we discover as we look across the valley yet another peak that calls. In the end, it is the wonderful experience of scaling the mountain—of attempting to understand the secrets of nature—that motivates us as scientists. There is of course the additional thrill, upon reaching the top of a mountain, to ram in the flagpole announcing one's victory. But that is only a momentary emotion soon superseded by the new challenges presented by the higher peak on the horizon.

I hope that non-physicists who have stuck with this book to the end have come away with a better understanding of how fundamental paradigms in physics are created, and the sociological and technical difficulties faced by those who attempt to shift established paradigms in science. I believe that with the ascent of each mountain—each theory, each paradigm—we reach a new truth, but that no mountain peak can ever represent the ultimate theory of nature. If you have gained an appreciation of that grandeur, and our attempts as physicists to describe it, from reading this book about gravity, then these three years of writing the book have been worthwhile.

For readers who have a background in physics, let us think about Einstein's gravitational theory in an abstract, whole sense. Then let us think about MOG. Do we get the same aesthetic pleasure from considering MOG as we do from Einstein's gravity theory? To reach a true appreciation of the elegance of MOG as a gravity theory, it is necessary to explore all the technical details of MOG and see how it works as a whole theoretical framework. One has to experience its successes in explaining data and naturally allowing for a cosmology with no singularity at the beginning of the universe, and no dark matter, and a unified description of the accelerating universe. Only after the laborious work of achieving a technical understanding of MOG's theoretical structure can one truly appreciate MOG's elegance. I hope that future generations of physicists will be motivated to study the theory in the same depth as Viktor Toth, Joel Brownstein, and I have done, and appreciate its intrinsic beauty.

There is still important research to perform before we have a complete picture of where we stand with modified gravity. Perhaps with more attention being paid by other physicists who can investigate MOG and apply it to other observational data, we will arrive at a more convincing state-of-the-art of modified gravity. The ultimate tests for MOG, or any alternative gravity theory, can be stated simply: With a minimum number of assumptions that are physically consistent, how much observational data can be explained? More important, can the theory make testable predictions that cannot be accounted for by competing theories? In the latter part of this book, I have suggested several ways that future observations and experiments can verify or falsify MOG, and discriminate between MOG and dark matter.

In probing the mysteries of nature, physicists need to have faith that we can through mathematical equations reach a true understanding of nature such that the predictions of our equations can be verified by experiment or observation. We need to continually aspire to that goal despite the modern trend in theoretical physics of indulging in speculations that can never be proved or falsified by reality.

Notes

INTRODUCTION

1. Quoted in Wali, p. 125.

CHAPTER 1: THE GREEKS TO NEWTON

1. Quoted in Gleick, p. 55.
2. Quoted in Christianson, p. 272.
3. Quoted in Gleick, p. 134.
4. Quoted in Boorstin, p. 407.

CHAPTER 2: EINSTEIN

1. Quoted in Pais, pp. 14–15.
2. Poincaré, pp. 94–95.
3. Einstein wrote three other famous papers in 1905:
    "A heuristic point of view concerning the production and transforma-
    tion of light," published in March 1905, describes the famous photoelectric
    effect. At the time, physicists faced the "ultraviolet catastrophe" due to the
    prevailing theory of blackbody radiation deviating from observation at high
    frequencies. Blackbody radiation refers to a physical system that absorbs all
    radiation and radiates energy in a spectrum that depends only on its tem-
    perature. This problem was solved by Max Planck in his celebrated paper on
    blackbody radiation in 1899. He quantized electromagnetic energy into small
    chunks, giving birth to the famous equation that states that energy equals
    Planck's constant h times the frequency of the electromagnetic radiation.

Until this time, physicists had believed that light only exists as continuous waves. Even Planck had difficulty accepting that radiation could sometimes take the form of discrete quanta of energy. Einstein believed that it was true, and in his paper explained that electrons would be ejected from certain metals when light strikes them. His interpretation of light fitted existing data. However, accurate experimental verification of the photoelectric effect had to await the experiments of the American physicist Thomas Millikan in 1922. Also, in that same year, the American physicist Arthur Compton proved that light acts as quanta when colliding with electrons and protons, and that data finally confirmed that Einstein was correct. The light quanta were eventually called photons. James Clerk Maxwell had discovered that light exists as electromagnetic waves. Einstein, in turn, discovered that light is also made of photons. This introduced the idea of the particle-wave duality of light.

"A new determination of molecular dimensions," which appeared in January 1906, was an adaptation of Einstein's PhD thesis of April 30, 1905. Although his PhD thesis was accepted by Zurich University, it was only twenty-one pages long, which the committee commented was too short. The thesis and the published paper dealt with the properties of particles suspended in a fluid. Einstein was able to calculate Avogadro's number (6.02 x 10^{23}, the number of atoms or molecules of any chemical substance in a gram mole), as well as the size of molecules by considering the motion of the particles in the fluid. This paper was the most cited during the years 1961 to 1975 for papers published prior to 1912.

"On the movement of small particles suspended in stationary liquids required by the molecular-kinetic theory of heat," published in May 1905, was Einstein's celebrated work on Brownian motion. Pollen grains jiggle around randomly in water due to the impact of millions of molecules. Einstein used kinetic theory and classical hydrodynamics to obtain an equation that showed that the displacement of the Brownian particles varies as the square root of the time t. The experimental vindication of this result came with the experiments by the French physicist Jean Perrin, proving once and for all the existence of atoms. Jean Perrin was awarded the Nobel Prize for physics for this work in 1926.

4. From E. Mach, *Die Mechanik in ihrer Entwicklung*, quoted in Weinberg, *Gravitation and Cosmology*, p. 16.

5. Newton also argued that if you tie two rocks together with a rope in empty space without matter, and you rotate the rope about its center, then the rocks will create an outward force pulling the rope taut. Again Newton noted that in an empty universe there is nothing against which to measure rotation other than absolute space. In the early development of Einstein's general relativity, there is no gravity in an empty universe, so special relativity becomes valid as a limiting case and all observers see the rock system rotating or accelerating. This argument tends to side with Newton and not Mach.

6. The concept of a metric tensor was developed in 1824 by Friedrich Gauss, the first mathematician to have the courage to accept non-Euclidean geometry.

 Gauss suggested a class of metric spaces such that for a sufficiently small region of space, it would be possible to find a local Euclidean coordinate system $[x_1, x_2]$ such that an infinitely close set of coordinates $[x_1 + dx_1, x_2 + dx_2]$ would satisfy Pythagoras's law: $ds^2 = dx_1^2 + dx_2^2$. When the space is not Euclidean, that is, when it represents the surface of a sphere or other curved surface and lines going to infinity are not parallel but can intersect one another, then the Euclidean-Pythagoric law must be replaced. It is replaced by an expression for ds^2 that contains coefficients associated with non-Euclidean metric measurements of distances between two points. These coefficients are described by a metric tensor.

7. In technical terms, Einstein had to develop a set of equations that would generalize the so-called Poisson equation. Developed by the famous French mathematician Siméon-Denis Poisson in 1813, this is a second-order differential equation for a scalar potential. On the right-hand side of the equation Poisson had Newton's gravitational constant times the density of matter. Einstein had to devise an equation that would reduce to Poisson's equation governing the law of gravity in Newton's theory for slowly moving bodies in weak gravitational fields. To this end, he devised an equation that had on the left side a contraction of Riemann's curvature tensor, called the Ricci tensor after its inventor, Gregorio Ricci-Curbastro, and on the right side Newton's gravitational constant divided by the fourth power of the speed of light times the energy momentum tensor. The form of this equation guarantees that the gravitational equations satisfy Einstein's principle of general covariance, namely, that the laws of physics remain the same when you transform from one reference frame to another.

8. At that time Einstein was unaware of the fact that a particular combination of the metric tensor with the Ricci curvature tensor contraction gave rise to four equations that were all satisfied identically, according to the generally covariant tensor equations of the gravitational theory. These would later be known as the Bianchi identities, named after the Italian mathematician who discovered them, Luigi Bianchi. Not knowing about these four identities was the basic reason that the gravity field equations Einstein published in 1913 violated the conservation of energy. Of the ten components in his symmetric metric tensor, four of them were arbitrary due to the general covariance of the theory, so only six components were to be determined from the six field equations that remained after the Bianchi identities were invoked.

9. Quoted in Pais, p. 253.

10. Hilbert employed the fact that the Hamiltonian function was invariant under general coordinate transformations and derived the correct form of the left-hand side of Einstein's field equations, including a crucial part involving the metric tensor multiplied by the scalar curvature tensor obtained by contracting the Ricci tensor. This left-hand side was equated to Newton's gravita-

tional constant divided by the fourth power of the speed of light times the energy momentum tensor. This constituted ten nonlinear partial differential equations and led correctly to the conservation of energy and momentum.

11. *The Times* (London), quoted in Isaacson, p. 261.

12. Quoted in Clark, p. 287.

CHAPTER 3: THE BEGINNINGS OF MODERN COSMOLOGY

1. Hoyle's cosmological solution was the same as de Sitter's "empty space" solution published in 1917 shortly after Einstein's paper on general relativity. The Hubble constant H was truly a constant of nature. Also, the pressure and density of matter in the universe were assumed to be constant in time. Hoyle found it necessary to modify Einstein's field equations, for with a constant pressure the conservation of energy would only be satisfied if the density were exactly equal to the negative value of the pressure in the universe divided by c^2. This required the introduction of Hoyle's C-field into the energy-momentum tensor of matter on the right-hand side of Einstein's field equations.

2. In 1952 the astrophysicist Edwin Salpeter suggested a two-step process. He proposed that two alpha particles (two helium nuclei) fused to form short-lived beryllium-8. If during that short lifetime a third alpha particle collided with the beryllium-8, then there was a small chance of creating a carbon-12 nucleus (six protons and six neutrons). But now a new problem arose. In order for the beryllium-8 and the alpha particle to form a nucleus of carbon-12, the combined mass-energies of these two components had to be very close to an energy state of carbon-12, possessing a similar mass-energy. Indeed the combined mass-energy must not exceed the mass-energy of the carbon-12 state, but must be a little smaller, to enable the energies of the particles moving around inside the hot star to bridge the gap of these two energy states. Such a state of carbon-12 is a "resonance," made of the beryllium-8 plus alpha particle system. When such a resonance state can be formed, it may then decay to the lowest energy or ground state of carbon-12 and create a stable carbon nucleus.

3. The two-step process that forms the resonance state at 7.6 million electron volts is a remarkable coincidence. The mass-energy of the state is larger than the combined masses of a beryllium-8 and an alpha particle, so that it allows for the decay of the resonant state into a beryllium-8 and an alpha particle. However, for every 10,000 decays, four result in the emission of gamma rays, which push the excited carbon-12 nucleus to its stable ground state. Although this is a rare occurrence, this result proved that the "carbon cycle" is the dominant mode of energy production in heavier stars.

CHAPTER 4: DARK MATTER

1. The concept of escape velocity is most easily understood in reference to the Earth. When a stone is thrown from the surface of the Earth with a speed v, it will reach a certain height when its kinetic energy is overcome by the Earth's gravitational force, and the stone will fall back to the ground. On the other hand, when engineers construct rockets that will go into orbit around the Earth or leave the vicinity of the Earth entirely, they ensure that the rocket's kinetic energy will be larger than the pull of Earth's gravity or the Earth's gravitational potential energy. The speed at which the rocket completely overcomes Earth's gravity is the escape velocity. The escape velocity of a test body will increase as the mass M of a gravitating body increases, and as the radius of the body decreases. Eventually, the gravitational pull of the larger object becomes so large that the escape velocity exceeds the speed of light, which leads to a black hole.

CHAPTER 5: CONVENTIONAL BLACK HOLES

1. However, if very short-distance quantum gravity effects or some strong, repulsive antigravity force caused, for example, by exotic matter or energy such as the dark energy claimed to be responsible for accelerating the expansion of the universe existed at the center of a collapsing star, then the destructive powers of the singularity could be quenched.

2. Physicists who dispute the claim that the supermassive object at the center of the galaxy is a black hole publish models in which the compact massive object has a surface that emits heat. This is in contrast to general relativity, which predicts that a black hole does not have a surface. Recently Avery Broderick and Ramesh Narayan, from the Harvard Smithsonian Astrophysical Institute, published a paper claiming to show that current observations do not favor the compact supermassive object having a surface, but rather an event horizon. The observed luminosity of the compact object of 10^{36} ergs per second implies an accretion rate of matter onto the gas surrounding the compact object that is a hundred times larger than what is expected from a blackbody heat-emitting stellarlike surface. Models show that a black hole event horizon can produce enough radiated heat to agree with the observed luminosity of the compact object. However, these calculations are based on critical assumptions about the astrophysical properties of the supermassive compact object. For example, the authors assume that the surface is in a steady state with respect to the accreting material, so that the surface radiates thermally like a pure blackbody, and that general relativity describes the gravity external to the surface. For such a supermassive compact object, these assumptions can be seriously questioned. Indeed for a black hole, the binding energy of the accreting matter increases the black hole's mass. Thus a black hole is an explicit example of an accreting compact object that cannot

be in a steady state. Moreover, the assumption of pure blackbody radiation is also suspect, for other forms of radiation might be expected from the extreme physical properties of a surface of such a compact, supermassive object. However, Broderick and Narayan argue that the extreme lensing of light by the massive object forces the emitted light back onto the object, creating a blackbody "shadow." As we shall see, modified gravity (MOG) can possibly produce gravity fields around a compact object that are not described by the Schwarzschild solution in general relativity and do not possess an event horizon or an essential singularity at the center.

CHAPTER 6: INFLATION AND VARIABLE SPEED OF LIGHT

1. In my 1993 published paper, I added what is called a Higgs mechanism associated with a vector field whose constant value in the vacuum does not vanish.

2. In a related paper written in 1992 and published in the journal *Foundations of Physics* in 1993, I also proposed a solution to the problem of time in general relativity that crops up when you attempt to unify gravitation theory with quantum mechanics. During the time when special relativity is spontaneously violated, near the Planck time of 10^{-43} seconds, time as a coordinate becomes like Newton's absolute time, allowing for the external, absolute time needed to describe quantum mechanics using Schrödinger's equation for the quantum mechanical wave function. Moreover, it also solved the old problem of the arrow of time, namely, why is there a direction of time that aligns itself with the increasing expansion of the universe? This question also leads to the need to explain how entropy always increases in accordance with the second law of thermodynamics.

3. The only prediction that could be tested observationally at present would explain the scale-invariant nature of the quantum fluctuations that form the seeds of galaxies in the cosmic microwave background. A critical test of inflation theory and VSL theory is the ratio of the gravitational wave fluctuations to the scalar matter mode fluctuations. This effect is hard to measure directly due to the lack of observation of gravitational waves, but an indirect measurement of the polarization of gravitational waves by future satellite observations could provide a critical test.

CHAPTER 7: NEW COSMOLOGICAL DATA

1. Quoted in Mather and Boslough, p. 252.

2. The Fourier analysis of a power spectrum is typically performed on a box-like object, whereas the power spectrum we see in the CMB is analyzed on a sphere, so we have to use the methods of spherical harmonics to determine the number of sine waves within the sphere. This technique is named after

the celebrated early nineteenth-century French mathematician Jean Baptiste Joseph Fourier, best known for initiating the investigation of mathematical Fourier series and their application to problems of heat flow and sound waves. Fourier analysis of acoustical waves tells, in the case of a violin, the shape of its sound chamber. In the case of the CMB, it effectively demonstrates the shape of the early universe.

3. Because the universe has been shown from the WMAP data to be spatially flat, we know that the density of the universe is equal to its critical density. This is the threshold density at which the universe can become either three-dimensionally closed, flat, or open. The universe is observed to be spatially flat and the visible baryon density is only 4 percent of the total matter and energy density needed to equal the critical density. Furthermore, the matter contained in stars is only about half a percent. Therefore in the standard model we are missing about 96 percent of the matter and energy in the universe.

From the Friedmann cosmological equations, we can determine a quantity called Omega (matter) (Ω_m). This quantity is the sum of all the constituents of matter and energy in the universe divided by the critical density. Ω_m determines the geometrical structure of the universe. When Ω_m is greater than one, the universe is spatially closed and finite in extent. When Ω_m equals one, the spatial curvature of the universe is zero, and the universe is called "flat." When Ω_m is less than one, the universe is spatially open. These correspond respectively to the values of the curvature constant $k > 0$, $k = 0$ or $k < 0$. Since the WMAP data have shown that the universe is spatially flat to within a 5 percent error, then we know that the curvature constant k is equal to zero, within the observational errors. Consequently, Ω_m equals one. However, as WMAP has observed, and from nuclear synthesis calculations, we have learned that the visible matter in the universe in the form of baryons only amounts to Ω_m equal to 0.04, or 4 percent of the total matter and energy budget needed to be consistent with the Friedmann equation and observations. So, to say it again, we are missing 96 percent of the matter and energy that makes the universe spatially flat.

There is another cosmological parameter called Ω that follows from the Friedmann equation, and represents the sum of Ω_m, Ω (curvature) (Ω_c) and Ω (Lambda) (Ω_Λ), where Ω_c denotes the fractional contribution of spatial curvature and Ω_Λ denotes the dark energy contribution as described by the cosmological constant Λ. The Friedmann equation tells us that this sum is equal to unity. Because the CMB data show that Ω_c is close to zero and that Ω_m is about 30 percent, then to get the total Ω sum to be unity, we are forced to conclude that there is a missing matter-energy component Ω_Λ that has to be about 70 percent. Therefore, we conclude that the CMB observations force us to include a dark energy contribution into the total matter-energy budget of the universe, in agreement with the supernovae observations of the accelerating expansion of the universe. The advocates of the standard cosmological model claim that this result, which follows from the combination of the CMB data and the supernovae data, is corroborating convincing evidence that the expansion of the universe is accelerating.

4. In my original paper on VSL, I derived a scale-invariant spectrum based on the quantum fluctuations of the field or particle responsible for spontaneously breaking the symmetry of special relativity. The calculation followed a similar route to that taken in inflationary models, but without the inflation of spacetime taking place. The critical part of the calculation was to allow the initial quantum fluctuations to penetrate the observable horizon and be frozen in the superhorizon part of the universe, thereby preventing the scale-invariant fluctuations from misbehaving and eventually breaking the scale invariance. Also, as the universe expands, these frozen imprints of the fluctuations could reenter the visible universe. The phase transition in the speed of light forced the horizon size to match the approximately constant value achieved in inflation theory for a short period of time, which allowed the fluctuations to get through to the superhorizon of the universe. These posthorizon fluctuations were causally connected by the superluminal speed of light in VSL theory.

5. One second to a minute after the big bang, the universe cooled down to a temperature equivalent to an energy of a million electron volts, low enough to allow nuclear synthesis to happen. This energy can be reproduced in nuclear physics laboratories, allowing physicists to see how the nuclear synthesis of the lighter elements actually happened. A proton and a neutron combine to produce a deuterium nucleus, accompanied by an emission of high-energy photons. Once deuterium is formed, we can fuse two deuterium nuclei to produce helium-3, consisting of two protons and one neutron plus one free neutron. During the big bang nuclear synthesis process, hydrogen and helium were abundantly formed. The helium consisted mainly of helium-4 (two protons and two neutrons), while isotopes were also produced such as deuterium, tritium (one proton and two neutrons) and helium-3. According to the measure of atomic mass, approximately a quarter of the nuclei in the universe are in the form of helium-4, with hydrogen making up almost all the rest of the matter content in the universe. Indeed, deuterium, tritium, helium-3 and lithium-6 and -7 nuclei should occur in much smaller but measurable amounts. In 1964, Fred Hoyle published the first calculation of the production of light elements by nuclear synthesis in the very early universe. Ironically, these calculations by one of the leading steady statesmen supported the idea of a big bang cosmology. Robert Wagoner, William Fowler, and Hoyle then performed a more complete calculation of the formation of the light elements. They also did calculations showing how the heavier elements were produced in stars. In the 1970s Wagoner produced a computer code that provided the most accurate nuclear synthesis calculations, which is still referred to today. Recently there has been controversy because the

predicted amounts of lithium-6 and -7 fall short of the measured values. The final word on nuclear synthesis is still not in.

6. Measurements of what is called the equation of state—the ratio of the vacuum pressure to the vacuum density—show that this ratio is −1 to within a few percent.

CHAPTER 8: STRINGS AND QUANTUM GRAVITY

1. The dimensions of Newton's gravitational constant G can be expressed as an inverse energy squared. When you multiply G as expressed in these dimensions by the square of an energy, and that energy is as large as the Planck energy, you obtain a number of order unity, which says that at the Planck energy gravity becomes a significant force of nature.

2. One reason for this failure was the problem that the pointlike photons and electrons, when made to interact with the hadronic strings, gave rise to long-range forces like the electromagnetic and gravitational forces. Photons and electrons do not have a size—they are infinitely small and cannot be spun like toy tops. Thus string theory had to require that electrons, photons, and gravitons also had to be strings in order to describe nuclear forces.

3. One of the problems in string theory from the very beginning has been the "moduli problem." The moduli are scalar fields (spin zero particles) that are huge in number and their shapes determine the way one compactifies the ten- or eleven-dimensional string theory and brane theory to four-dimensional spacetime—stuffing all those extra clothes into the tiny suitcase of reality. The goal of compactification is to find a low-energy theory that describes the very successful standard model of particle physics. The compactification of the ten-dimensional string theory should also lead to a vacuum state that is the ground state of minimum energy for the standard model of particles.

Extensive studies of these compactifications of the higher-dimensional theory by Shamit Kachru, Renata Kallosh, Andrei Linde, and Sandip Trivedi at Stanford University came to be known as the "KKLT" description of the valleys and peaks of the string theory landscape; the main paper was published in the *Physical Review* in 2003. The KKLT study revealed two results: 1) all the vacuum states were metastable states, that is, they only were stable for a short amount of time that had to be fine-tuned according to the age of the universe, and 2) the number of such solutions of string theory with vacuum states again turned out to be huge, and possibly infinite. The number of possible solutions has been quoted as being of the order 10^{500}, plus or minus a few trillion! Another problem is that the KKLT study does not appear to have revealed a solution with a cosmological constant of the right size to agree with cosmological observations. Leonard Susskind has derogatorily dubbed the KKLT construction a "Rube Goldberg device."

4. This new idea of a multiverse is reminiscent of the old parallel universes hypothesis, proposed by Hugh Everett III in the mid-1950s. He called his interpretation of quantum mechanics the "many-worlds" interpretation. His purpose was to explain the probability basis of quantum mechanics. Everett's idea is that each time a probability measurement is made, for example, about the position and momentum of a particle, then each potential outcome that can occur takes place in a new universe. Thus the quantum mechanical states of the universe inhabited by the particle are like the pages in a huge encyclopedia of the universe. Each page describes one possible history of the particle. When a measurement is made of the position and momentum of the particle, then only one page of this many-worlds encyclopedia is chosen out of the many possible pages. Because quantum mechanics is probabilistic rather than deterministic, we are inevitably led to a many-histories interpretation. This modern interpretation of quantum mechanics has been advocated by physicists such as John Wheeler, Richard Feynman, Murray Gell-Mann, Brandon Carter, and James Hartle, among others.

Some physicists have extended the Everett many-worlds interpretation to the idea of a multiverse. As in the case of Everett's many-worlds interpretation, we are faced with the fact that we can only observe our universe—we only have one universe—and so, we end up in a philosophical quagmire. A technique that is related to the multiverse is called the "path integral formulation" in quantum mechanics. It was pioneered in 1948 by Richard Feynman. He worked out some of the preliminary features in his doctoral thesis with his supervisor, John Wheeler. By the 1970s, the path integral formulation proved to be a grand synthesis of quantum field theory that unified renormalization theory, leading to finite calculations in quantum field theory and statistical mechanics. Feynman was initially inspired by comments made by Paul Dirac about the quantum equivalence of the classical action principle based on the Euler-Lagrange equations. The idea is that a quantum amplitude consists of all the possible paths that a particle moving from point A at some time t_0 to a point B at a later time t_1. Feynman proposed the following principles:

> The probability for an event to happen in quantum mechanics is given by the square of a complex-number valued amplitude.
> The quantum amplitude for an event is given by adding up all the histories which lead to the event.
> The amplitude given by the total sum of histories is proportional to the exponential of the square root of –1 times the action S divided by Planck's constant h, where the action S is the time integral of the Lagrangian of the quantum theory.

Feynman's formulation of quantum mechanics has a teleological quality: One starts with a combination of initial conditions and final conditions, and then finds the actual path taken by a particle from all the probable paths that the particle could choose. Somehow the particle does not know where it is going but does in the end make a definite decision through an observable measurement when we calculate the path integral probability for a given physical process. Technically speaking, the path integral determines the probability amplitude for a physical process, and stationary points in the action determine neighborhoods of the space of histories whose quantum mechanical interference yields large probabilities for the final outcome.

Feynman's path integral formalism is related at a deep level to the many-world's interpretation of quantum mechanics, in that all the possible paths that a particle can travel between two points can be interpreted as happening in parallel universes that branch out at the initial starting point and encompass a sum of all the possible universes for each individual path taken by the particle. Again quantum mechanics leads us to a multiverse picture of reality. But again this interpretation of quantum mechanics, due to its strange random probability, cannot be distinguished by any known experiment from the conventional Copenhagen interpretation of quantum mechanics that uses the notion of a collapsing wave function to determine the outcome of a measurement by an external observer's apparatus.

5. Wilson loops can describe a gauge theory such as Maxwell's theory of electromagnetism or the gauge theory of the standard model of particle physics. These loops are gauge-invariant observables obtained from the holonomy of the gauge connection around a given loop. The holonomy of a connection in differential geometry on a smooth manifold is defined as the measure to which parallel transport around closed loops fails to preserve the geometrical data being transported. Holonomy has nontrivial local and global features for curved connections.

Wilson loops were introduced in an attempt at a nonperturbative formulation of quantum chromodynamics (QCD), the theory based on quarks of the strong force that binds the atomic nucleus. The Wilson loops were introduced to solve the problem of the confinement of quarks inside protons and neutrons and other basic hadrons. The confinement problem remains unsolved to this day. Interestingly, the strongly-coupled quantum field theories have elementary nonperturbative excitations which are loops, and they motivated Alexander Polyakov at Princeton University to develop the first string theories that describe the propagation of elementary quantum loops in spacetime. In loop quantum gravity, the Wilson loops are made from the gauge connections of formulations of general relativity in terms of geometrical gauge fields as published by Abhay Ashtekar, Carlo Rovelli, and Lee Smolin.

6. From the theory called quantum chromodynamics, in addition to the normal quark charges we also have "color charge." This is equivalent to the coupling strength responsible for confining the quarks inside the nucleon, making it impossible for us to observe them.

7. A leader in the search for noncommutative quantum gravity is the French mathematical physicist Allan Connes. The possibility of picturing spacetime as a quantum mechanical noncommuting structure was first investigated by Harlan Snyder, who was Robert Oppenheimer's student at the University of California at Berkeley in the late 1930s; the two of them published the first solution to the gravitational collapse of a star to a black hole in Einstein's general relativity.

8. In 2000, I published a paper in *Physics Letters B* showing that if one applies the noncommuting ideas to Einstein's gravity theory, then the metric tensor becomes a complex quantity, that is, it has to be described in terms of both real and imaginary numbers involving the square root of minus one. The complex nature of the metric tensor comes about by the use of Moyal-star products of vierbeins (German for "four legs"): four vector quantities whose products describe the spacetime metric. This is a serious departure from standard general relativity and requires a significant modification of its geometrical structure in ways that are not fully understood at present. In my first published paper, in the *Proceedings of the Cambridge Philosophical Society*, when I was a student at Trinity College in the 1950s, I formulated a complex symmetric Riemannian geometry, and in a subsequent paper, I generalized Einstein's general relativity to a complex symmetric Riemannian differential geometry. This complex geometry could perhaps underlie the noncommutative spacetime quantum gravity.

Moreover, as I showed in a later paper in *Physics Letters B*, ignoring the complication of basing gravity on a complex metric and spacetime, and applying standard perturbative calculational techniques to the noncommutative quantum gravity, leads to more or less the same infinities that plague quantum gravity calculations using the very successful techniques employed in modern quantum field theory. I found this result a serious shortcoming of the noncommutative spacetime approach to quantum gravity.

In 1990, when I was a professor at the University of Toronto, I entered the quantum gravity arena with another approach. I published a paper in *Physical Review D* proposing that quantum field theory be modified to be a nonlocal quantum field theory. In the standard modern quantum field theory, one of the basic axioms is that measurements of quantum fields at short distances satisfy causality, that is, the effect of an event in spacetime follows chronologically after its cause. The measurement of a quantum field at a point in space is only *locally* affected by another measurement at a different point in space, and measurements of fields at points far away from the local measurements do not affect the physics locally. Of course the *nonlocality* of measurements of fields has the opposite meaning, that is, measurements made at distant points in space can causally affect measurements in the local neighborhood of the measuring apparatus. By giving up the physical requirement of causality at the short distances encountered in particle physics—that is, at a scale of less than 10^{-14} centimeters—it is possible to obtain a quantum field theory that does not have the tenacious infinities that keep appearing in standard quantum field theory and quantum gravity.

In my paper I proposed a way to maintain the basic gauge symmetries of quantum field theory essential for the consistency of the theory, and to keep out negative energy "ghosts." Moreover, it was also possible to maintain the axiom of unitarity of quantum scattering amplitudes, which describes the scattering of elementary particles off one another—that is, that probabilities always remain positive, a clearly necessary requirement in physics. I developed a general program for making a finite, consistent nonlocal field theory that is also a quantum gravity theory. At the time, I used only Einstein's gravity theory as the basis of the quantum gravity theory, although I did suggest that one could use a MOG if future experiments demanded that MOG replace Einstein's gravity theory. In 1991, I published a more detailed version of this quantum field theory in *Physical Review D* with collaborators Richard Woodard, Dan Evens, and Gary Kleppe. We provided a detailed proof of the consistency of the nonlocal field theory formalism in quantum electrodynamics. Further developments of the nonlocal field theory were published by myself and collaborators and Kleppe and Woodard.

Standard quantum mechanics already incorporates the idea of nonlocality. The Irish physicist John Bell discovered this to be true. His famous theorem follows from the notion of random probabilities in quantum mechanics and has been confirmed by experiment. This nonlocal phenomenon in quantum mechanics reveals itself dramatically in the "quantum entanglement" of distant photons. The connection between the nonlocality of the nonlocal quantum mechanics and nonlocal quantum field theory and quantum gravity is not yet well understood.

One advantage of this program for obtaining a finite quantum gravity theory is that it succeeds in deriving finite calculations while retaining the basic symmetries of gravity theories, such as local Lorentz invariance and general covariance, while the formalism takes place in four spacetime dimensions and not the ten dimensions of superstring theory. You do not have to go to ten dimensions to get a consistent quantum gravity theory, as is required in superstring-brane theories. Also the calculations are done with one vacuum state as in the standard model of particle physics, and not 10^{500} vacuum states as in the landscape of string theory. String field theory is already a nonlocal theory, so the goals of string theory, as far as formulating a finite quantum gravity theory are concerned, are already contained in the nonlocal quantum gravity theory. However, it shares with current string theory the feature that the calculations of quantum amplitudes are done on a fixed geometrical background, for example, a flat Minkowskian background. This is criticized by loop quantum gravity proponents, who insist on a quantum gravity theory being independent of the background. Indeed, this independence of background would seem to be a desirable feature of a quantum gravity theory, but I think that from a practical point of view, the derivation of a self-consistent, finite quantum gravity theory may not depend significantly on whether it is background independent or not. It is possible that we can make the nonlocal quantum gravity theory in four spacetime dimensions background independent, but no one has successfully achieved this yet.

At the time we published our papers on nonlocal field theory and quantum gravity, there was a new surge of interest in string theory due to the so-called second revolution in string theory with the discovery of M-theory. We made some mild critical comments of string theory, in the introduction of our paper in 1991, and this did not go down well with string theorists. Due to the sociological and media forces that drive much of modern physics, our papers on this subject have not received the attention that I believe they should.

CHAPTER 9: OTHER ALTERNATIVE GRAVITY THEORIES

1. More recent attempts to understand the weakness of the gravitational force have been made using higher-dimensional theories with strings and branes, in which the gravitational force exists in a five-dimensional "bulk" between two three-dimensional branes. Two prominent authors who have proposed such a theory are Lisa Randall at Harvard University and Raman Sundrum at Johns Hopkins University. Modern string theory is also a kind of modified gravity theory. But its nine spatial dimensions and one time dimension do not reduce to pure Einstein gravity in four dimensions on compactification of the other six dimensions. Instead, it reduces to a Jordan-Brans-Dicke-type modified gravity theory with many possible scalar fields. These scalar fields are called "modular fields" and play an important role in string theory when reducing the ten dimensions of string theory to four dimensions.

2. This theory involves fourth-order differential equations, which can lead to instability problems, and the conformal nature of the field equations leads to matter having effectively no mass. This is allowable in Maxwell's electromagnetism, which is physically a conformally invariant theory, in which angles but not directions in space are preserved, because we know the photon has no mass. The gravitational field equations possess a matter content that exerts the force of gravity on other bodies or, equivalently, warps the geometry of spacetime. Einstein's theory of gravity is not conformally invariant because the mathematical trace of the energy momentum tensor does not vanish, whereas the equivalent trace of the energy momentum tensor in Maxwell's theory does vanish because the photon is massless. To overcome this problem, Mannheim proposed to break the conformal invariance of his theory, producing the observed matter density of the universe. Recently, in collaboration with Carl Bender of Washington University in St. Louis, he has developed methods to avoid instabilities in his conformal gravity theory.

CHAPTER 10: MODIFIED GRAVITY (MOG)

1. I began by making the weak field approximation equations in NGT part of a Riemannian geometry. The field equations were covariant under general coordinate transformations. I had discovered a modification of Einstein's gravitational theory with a new degree of freedom described by an antisymmetric, second-rank tensor field, which by the mathematical operation of the "curl" generated a third-rank antisymmetric field strength. This tensor field theory without the inclusion of gravity was originally published by Pierre Ramond and his student Michael Kalb in the *Physical Review* in 1974. The mathematical structure of my new gravity theory was certainly simpler than NGT and it reduced to Einstein's general relativity when the Kalb-Ramond field vanished. In this way it contained all the predictions in agreement with experimental data in Einstein's gravity theory. Since the theory was based on an Einstein symmetric metric and an antisymmetric (or skew symmetric) field, I called the theory Metric-Skew-Tensor Gravity (MSTG).

2. I was also able to construct a classical action principle whose variation according to the least action principle of Maupertuis, Lagrange, and Hamilton led to a set of field equations for gravity and the new fields. The field equations for the vector phion field were analogous to the field equations for Maxwell's electromagnetic fields except that the phion particle replaced the photon and it had a mass.

3. The acceleration law had the form of the Newtonian acceleration law for a test particle plus an additional repulsive acceleration associated with what is called a "Yukawa potential." This potential was discovered by Hideki Yukawa in the 1930s in the process of investigating the nuclear forces. Using this potential, he predicted the existence of the meson, a subatomic particle, and was awarded the Nobel Prize in 1949 after the meson was found. In STVG, the mass of the vector phion field entered into the Yukawa potential and determined the distance range parameter λ (lambda) of the potential. When λ is infinitely large, the potential reduces to the inverse square law of Newtonian gravity. For a finite λ, the Yukawa potential vanishes exponentially fast as the distance from the source increases; it reduces to an inverse square law potential when the distance from the source is much less than λ.

   In an early attempt at modifying the laws of gravity to describe the galaxy rotation curve data without dark matter, Robert Sanders had in 1986 suggested using a Yukawa potential combined with Newton's potential. However, his published paper was based on a purely nonrelativistic, phenomenological proposal, whereas my modified Newtonian law was derived from a fully relativistic theory of gravitation. Also, his modification could not lead to agreement with the solar system observations and did not correctly describe a large amount of galaxy rotation data and data from hot X-ray clusters of galaxies.

4. http://cosmicvariance.com/2006/08/21/dark-matter-exists/

CHAPTER 12: MOG AS A PREDICTIVE THEORY

1. Solving such field equations involves obtaining solutions from differential equations that contain constants of integration. These constants of integration are fixed by using accurate galaxy rotation curve data, and once fixed, the modified Newtonian acceleration law and the relativistic equations for lensing and cosmology can be universally used to predict all the applications to observational data from the solar system to the cosmological horizon of the universe. Thus we find that MOG has no free adjustable parameters. MOG now has the form of a classical field theory in which the solutions of the field equations and all their applications to observational data can be obtained from the theory's basic action principle. This imbues MOG with significant predictive power.

CHAPTER 13: COSMOLOGY WITHOUT DARK MATTER

1. There were other problems to solve in developing MOG to the point that it could fit the data as well as Einstein plus dark matter. One problem lurked in the background when one amplified the 4 percent baryon density to about 30 percent by increasing the size of Newton's constant G: The baryons in the hot baryon-photon fluid plasma are tightly coupled to the photons through electromagnetic interaction. This means that the pressure of the photons exerts a drag force on the baryons, which has the effect of damping the acoustical waves observed in the WMAP data. This drag force was first discovered by the Oxford cosmologist Joseph Silk in the early 1960s, and it can erase the oscillation peaks in the acoustical power spectrum. However in MOG, the increased strength of the gravitational constant before decoupling reduces this drag force. This is enough to reduce the suppression of the acoustical waves so that MOG can fit the third peak in the acoustical power spectrum. In the standard model including dark matter, we recall that the dark matter does not couple to the photons or the baryons and therefore Silk damping is not a problem.

2. In our application of MOG, we simplified the field equations by working in a weak gravitational field that would occur at large cosmological distances. We used the MOG acceleration law, which includes a varying G, and a repulsive short-range fifth force due to the phion field. At large distances from a gravitating body with a given mass, this modified weak gravity theory becomes Newton's theory of gravitation but with a different gravitational constant. In a homogeneous and isotropic FLRW universe, there will not be any distance-dependent modification of Newtonian gravity. However in MOG, the gravitational constant is not Newton's constant, but the previously mentioned constant measured at very large distances from an observer on Earth. As we move from the very early universe, which is homogeneous and isotropic,

toward the more recent universe when stars and galaxies have been formed and the universe is no longer homogeneous, then the varying G can show a distance dependence. This then leads to a prediction for the growth of the fluctuations that are first seen at the surface of last scattering in the CMB data, and a prediction of this growth in the matter power spectrum discussed in Chapter 7.

3. In practice, when we calculate the matter power spectrum using the standard model, we have to use what is called a "transfer function." This function was first published by Wayne Hu and Daniel Eisenstein in 1998 in *Physical Review*. It consists of two parts. One part is due to the baryons, and the other part is due to the cold dark matter. You take the square of this transfer function, and this produces the power spectrum as a function of the wave number k. As it turns out, the baryon part of the transfer function is proportional to a Bessel function, first discovered by the German mathematical physicist Friedrich Bessel in 1817. This function oscillates like sine waves. The second cold dark matter part of the transfer function is not proportional to any oscillating function, and produces a smooth contribution to the power spectrum. When you use Einstein gravity and just Newton's gravitational constant, then the cold dark matter part of the transfer function dominates the matter power spectrum and produces the smooth Lambda CDM fit to the data. In contrast, MOG, through the oscillating Bessel transfer function, produces an oscillating curve that fits the matter power spectrum data (after having used a window function in the calculation, which takes into account the size of the errors of the calculation of the power spectrum and the errors in the power spectrum data). Therein lies the crucial difference between modified gravity based purely on baryon matter and the Lambda CDM model based on dominant cold dark matter.

CHAPTER 14: DO BLACK HOLES EXIST IN NATURE?

1. A. Einstein, "Autobiographical Notes," in Schilpp, p. 95.

CHAPTER 15: DARK ENERGY AND THE ACCELERATING UNIVERSE

1. As the cosmologist George Ellis proposed in the mid-1980s when investigating inhomogeneous solutions of Einstein's field equations, it is necessary to perform a spatial averaging of the matter density and curvature of the universe. This averaging must take into account the nonlinear nature of Einstein's field equations, whereby the rate of change of an average value of a physical quantity (for example, the expansion of the universe) is not equal to the *average of the rate of change* of this quantity. By correctly perform-

ing this averaging, one can avoid the no-go theorem of the Raychaudhuri equation, which says that you cannot have an accelerating expansion of the universe locally if the density and pressure are positive.

2. Quoted in Krauss, p. 8.

CHAPTER 16: THE ETERNAL UNIVERSE

1. "Un Ora," *Acta Apostolicae Sedes—Commentarium Officiale*, 44 (1952): 31–43. Quoted in Farrell, p. 196.

2. The second law of thermodynamics is closely linked with the arrow of time and the theory of thermodynamics, which began with the industrial revolution in Britain and the Scotsman James Watt, who constructed the first steam engine. The French engineer Sadi Carnot developed in the early 1800s a theory of thermodynamics by considering a perfectly reversible engine with no irreversible heat loss. However, he realized that there is no perfect heat engine—the heat loss could never reach zero. The Englishman James Prescott Joule understood the equivalence of heat and work, and demonstrated that work produced an equal amount of heat and had a unit of energy, the Joule, named after him.

This equivalence of heat and work led to the first law of thermodynamics, amounting to the conservation of energy in a physical system. Energy gets converted from one form to another without disrupting the balance sheet of energy, although a dissipation of energy does occur in physical processes. The German physicist Rudolf Clausius, in 1850, thought about this dissipation of energy and realized that when work gets converted to heat, some of it always gets lost and cannot be recovered. He had discovered the arrow of time. William Thomson, later Lord Kelvin, enunciated the physical law now known as the second law of thermodynamics.

In 1865, Clausius introduced the idea of entropy from the Greek words *en*, meaning "in," and *trop*, meaning "turning." Carnot had stated that heat only flows from a hot to a cold body, and this immediately implies an arrow of time. When entropy reaches a maximum value, then the system is in its most disordered and random state. The German physicist Hermann von Helmholtz went further and applied the second law of thermodynamics to the cosmology of the universe. The final state of the universe due to the inexorable increase in entropy would be a universe in equilibrium—no more change would take place and disorder would be at a maximum: the heat death. However, it was later realized that gravity would have to be overcome, a circumstance now made pertinent by the discovery that the expansion of the universe is accelerating.

The second law of thermodynamics has been subjected to much controversy. Because of the time symmetry of the fundamental laws of physics, such as Newtonian mechanics, it took the brilliance of the German physicist

Ludwig Boltzmann to solve the problem of how to extract an asymmetric temporal evolution from a time symmetric assembly of molecules and atoms. He discovered what is called the "Boltzmann equation," satisfied by a distribution function—a statistical description of the motion of a single molecule in a gas. His equation violated the time-symmetric Newtonian law of motion.

Boltzmann was vigorously attacked by contemporary physicists such as Josef Loschmidt. Henri Poincaré's recurrence theorem was used to criticize Boltzmann's purported derivation of the arrow of time. This theorem, stated by the celebrated French mathematician, says that every isolated physical system must return to its original state. Boltzmann was already heavily criticized by Ernst Mach and Wilhelm Ostwald, among others, for his interpretation of thermodynamics as an atomic and molecular phenomenon. It is speculated that this caused him to fall into a depression and to commit suicide in 1906.

In recent years the arrow of time has been dismissed by physicists and philosophers as a subjective concept. A reinterpretation of statistical mechanics and thermodynamics using the concept of *coarse-graining* has gained some popularity. This says that we have to reinterpret Boltzmann's equation. At microscopic distances we have to calculate the average motions of atoms in finite subcells of a gas. This suggests that Boltzmann's thermodynamic arrow of time is illusory. However, invoking the *course-graining* technique is somewhat ad hoc. The introduction of the modern theory of information has linked the communication of information with entropy.

In formulating my cosmology with a start to the universe at t = 0, I have strictly adhered to the second law of thermodynamics and an asymmetric arrow of time running in both the positive and negative directions away from t = 0. I take to heart Arthur Eddington's comment in *The Nature of the Physical World*: "I wish I could convey to you the amazing power of the concept of entropy in scientific research."

3. About 130 years ago, the inventor of the statistical theory of thermodynamics, Ludwig Boltzmann, asked the question about the universe: How can we reconcile the second law of thermodynamics with the observed asymmetry of time? This prompted the related question: Since entropy always increases, how low was it in the first place? He proposed that the universe started as a chance fluctuation in a state of minimal entropy. Although such a chance fluctuation was extremely unlikely in his statistical theory of thermodynamics, nonetheless, in a universe that lasts a very long time it could indeed occur.

Many physicists since Boltzmann have sought to explain temporal asymmetry and the place of the second law of thermodynamics in cosmology. Most notably Thomas Gold, Stephen Hawking, Roger Penrose, and Paul Davies have attempted to reconcile the big bang model and other cosmological models with the dual conundrum of temporal asymmetry and the increase of disorder from a singular beginning in Einstein's gravity theory. If we entertain the idea that the entropy is low at the big bang (when paradoxically we might expect it to be at a maximum or even infinite), then we must face the perplexing question of why it will not be equally low at the big crunch, if we contemplate a collapse of the universe to a singularity in the future. Unless we entertain with Gold and Hawking that the order of temporal events reverses before we reach the big crunch, together with a reversal of the thermodynamic arrow of time, then we succumb to a violation of the second law of thermodynamics. A reversal of temporal events would mean that the light from stars would reverse and leave our eyes and be absorbed by stars, and apples fallen from a tree would bounce back into the tree. However, biological and thought processes would still operate in humans, so we would be unaware of the "unnatural" state of affairs in the backward-in-time universe.

The meaning of a finite universe with a "start" as opposed to an infinite lasting universe has also been a topic of considerable debate over the ages with significant implications for religion and philosophy. Immanuel Kant posed the paradox in 1781 in his *Critique of Pure Reason*, that if the universe had a beginning at a finite time in the past, what came before it? And what caused it to happen before it happened, and so on. He thought that an infinite universe would be equally unacceptable.

Our laws of physics such as Newtonian mechanics, Maxwell's equations of electromagnetism, Einstein's gravitational equations and the Schrödinger equation are surprisingly symmetric in time. They do not distinguish past from future, so it is far from obvious how to derive the observed temporal asymmetry and the thermodynamic arrow of time—disorder continually increases—from our accepted laws of physics.

I have attempted to develop a MOG cosmology that is shaped by a singularity-free beginning at to the universe and a strict adherence to the second law of thermodynamics. This is accomplished by imposing boundary conditions on the MOG field equation solutions such that the universe truly starts with no matter and energy at time $t = 0$, an empty universe with zero gravity and minimal entropy. Fluctuations in the vacuum at $t = 0$ and the production of particles from the vacuum create matter and the curvature of spacetime, causing the universe to expand in both the positive and negative directions of time with increasing entropy and disorder and with temporal asymmetry. This is accomplished by a reversal of temporal events when comparing the dynamical expansions of the universes starting at $t = 0$ and evolving in positive and negative time. For those in either universe there is only a past and a future; there is no possibility of going back to the past; time machines are prevented from existing. There is no violation of causality in either evolving universe.

We know experimentally that the invariance of the symmetry of time is violated in K-meson decay. The weak decay of a neutral K-meson to a pi-meson, an electron and an antineutrino violates what is known as CP invariance; this was discovered in 1964 by Christensen, Cronin, Fitch, and Turlay. Here C is the conjugation of charge and P is the (parity) reflection of the spatial coordinate. In 1957 it was discovered that the weak interaction violated parity invariance P. Therefore it came as a big surprise to find that the combination of charge conjugation C, which changes the sign of the electric charge, and the parity P—namely, CP—was also violated. It can be shown that the combined operation of CPT (assuming global Lorentz invariance) must be invariant in particle physics. Therefore the basic law of time reversal invariance T must also be violated, contradicting this symmetry in our basic laws of physics. This constitutes a significant violation of the symmetry of time. However, even though this is the only microscopic manifestation of this violation of time reversal invariance, it does not affect the second law of thermodynamics.

There is at present no satisfactory explanation for the CP violation in the weak decay of neutral K-mesons or for the more recently observed weak decays of B-mesons. However, there has been speculation, notably by the French physicist Gabriel Chardin, that "antigravity" may be responsible for the violation of CP symmetry and its accompanying violation of time reversal invariance T. This invites the speculation that this antigravity could be caused by the repulsive force in MOG produced by the phion vector field.

Although our MOG cosmology solves the dual problems of the increase of disorder and entropy and the observed asymmetry of time, it can invite the criticism that I have imposed an initial condition on the universe. The universe starts in either positive or negative directions in time from $t = 0$ in a minimal entropy state (maximum order) and always expands to a maximum disordered state (maximum entropy) in the infinite future with a heat death or a cold and empty universe.

We can counter the criticism that I have imposed an initial condition on the universe by acknowledging that fundamental equations of physics are solved with imposed boundary conditions. The wave solutions of Maxwell's equations allow both advanced and retarded solutions. We demand that only the retarded solutions are allowed to describe electromagnetic wave motion, for they do not violate causality. We must not expect to obtain a cosmology describing the whole evolution of the universe from a theory such as MOG without imposing a special initial condition on the solutions of the field equations.

Glossary

Absolute space and time—the Newtonian concepts of space and time, in which space is independent of the material bodies within it, and time flows at the same rate throughout the universe without regard to the locations of different observers and their experience of "now."

Acceleration—the rate at which the speed or velocity of a body changes.

Accelerating universe—the discovery in 1998, through data from very distant supernovae, that the expansion of the universe in the wake of the big bang is not slowing down, but is actually speeding up at this point in its history; groups of astronomers in California and Australia independently discovered that the light from the supernovae appears dimmer than would be expected if the universe were slowing down.

Action—the mathematical expression used to describe a physical system by requiring only the knowledge of the initial and final states of the system; the values of the physical variables at all intermediate states are determined by minimizing the action.

Anthropic principle—the idea that our existence in the universe imposes constraints on its properties; an extreme version claims that we owe our existence to this principle.

Asymptotic freedom (or safety)—a property of quantum field theory in which the strength of the coupling between elementary particles vanishes with increasing energy and/or decreasing distance, such that the elementary particles approach free particles with no external forces acting on them; moreover for decreasing energy and/or increasing distance between the particles, the strength of the particle force increases indefinitely.

Baryon—a subatomic particle composed of three quarks, such as the proton and neutron.

Big bang theory—the theory that the universe began with a violent explosion of spacetime, and that matter and energy originated from an infinitely small and dense point.

Big crunch—similar to the big bang, this idea postulates an end to the universe in a singularity.

Binary stars—a common astrophysical system in which two stars rotate around each other; also called a "double star."

Blackbody—a physical system that absorbs all radiation that hits it, and emits characteristic radiation energy depending upon temperature; the concept of blackbodies is useful, among other things, in learning the temperature of stars.

Black hole—created when a dying star collapses to a singular point, concealed by an "event horizon;" the black hole is so dense and has such strong gravity that nothing, including light, can escape it; black holes are predicted by general relativity, and though they cannot be "seen," several have been inferred from astronomical observations of binary stars and massive collapsed stars at the centers of galaxies.

Boson—a particle with integer spin, such as photons, mesons, and gravitons, which carries the forces between fermions.

Brane—shortened from "membrane," a higher-dimensional extension of a one-dimensional string.

Cassini spacecraft—NASA mission to Saturn, launched in 1997, that in addition to making detailed studies of Saturn and its moons, determined a bound on the variations of Newton's gravitational constant with time.

Causality—the concept that every event has in its past events that caused it, but no event can play a role in causing events in its past.

Classical theory—a physical theory, such as Newton's gravity theory or Einstein's general relativity, that is concerned with the macroscopic universe, as opposed to theories concerning events at the submicroscopic level such as quantum mechanics and the standard model of particle physics.

Copernican revolution—the paradigm shift begun by Nicolaus Copernicus in the early sixteenth century, when he identified the sun, rather than the Earth, as the center of the known universe.

Cosmic microwave background (CMB)—the first significant evidence for the big bang theory; initially found in 1964 and studied further by NASA teams in 1989 and the early 2000s, the CMB is a smooth signature of microwaves everywhere in the sky, representing the "afterglow"of the big bang: Infrared light produced about 400,000 years after the big bang had redshifted through the stretching of spacetime during fourteen billion years of expansion to the microwave part of the electromagnetic spectrum, revealing a great deal of information about the early universe.

Cosmological constant—a mathematical term that Einstein inserted into his gravity field equations in 1917 to keep the universe static and eternal; although he later regretted this and called it his "biggest blunder," cosmologists today still use the cosmological constant, and some equate it with the mysterious dark energy.

Coupling constant—a term that indicates the strength of an interaction between particles or fields; electric charge and Newton's gravitational constant are coupling constants.

Crystalline spheres—concentric transparent spheres in ancient Greek cosmology that held the moon, sun, planets, and stars in place and made them revolve around the Earth; they were part of the western conception of the universe until the Renaissance.

Curvature—the deviation from a Euclidean spacetime due to the warping of the geometry by massive bodies.

Dark energy—a mysterious form of energy that has been associated with negative pressure vacuum energy and Einstein's cosmological constant; it is hypothesized to explain the data on the accelerating expansion of the universe; according to the standard model, the dark energy, which is spread uniformly throughout the universe, makes up about 70 percent of the total mass and energy content of the universe.

Dark matter—invisible, not-yet-detected, unknown particles of matter, representing about 30 percent of the total mass of matter according to the standard model; its presence is necessary if Newton's and Einstein's gravity theories are to fit data from galaxies, clusters of galaxies, and cosmology; together, dark matter and dark energy mean that 96 percent of the matter and energy in the universe is invisible.

Deferent—in the ancient Ptolemaic concept of the universe, a large circle representing the orbit of a planet around the Earth.

Doppler principle or **Doppler effect**—the discovery by the nineteenth-century Austrian scientist Christian Doppler that when sound or light waves are moving toward an observer, the apparent frequency of the waves will be shortened, while if they are moving away from an observer, they will be lengthened; in astronomy this means that the light emitted by galaxies moving away from us is redshifted, and that from nearby galaxies moving toward us is blueshifted.

Dwarf galaxy—a small galaxy (containing several billion stars) orbiting a larger galaxy; the Milky Way has over a dozen dwarf galaxies as companions, including the Large Magellanic Cloud and Small Magellanic Cloud.

Dynamics—the physics of matter in motion.

Electromagnetism—the unified force of electricity and magnetism, discovered to be the same phenomenon by Michael Faraday and James Clerk Maxwell in the nineteenth century.

Electromagnetic radiation—a term for wave motion of electromagnetic fields which propagate with the speed of light—300,000 kilometers per second—and differ only in wavelength; this includes visible light, ultraviolet light, infrared radiation, X-rays, gamma rays, and radio waves.

Electron—an elementary particle carrying negative charge that orbits the nucleus of an atom.

Eötvös experiments—torsion balance experiments performed by Hungarian Count Roland von Eötvös in the late nineteenth and early twentieth centuries that showed that inertial and gravitational mass were the same to one part in 10^{11}; this was a more accurate determination of the equivalence principle than results achieved by Isaac Newton and, later, Friedrich Wilhelm Bessel.

Epicycle—in the Ptolemaic universe, a pattern of small circles traced out by a planet at the edge of its "deferent" as it orbited the Earth; this was how the Greeks accounted for the apparent retrograde motions of the planets.

Equivalence principle—the phenomenon first noted by Galileo that bodies falling in a gravitational field fall at the same rate, independent of their weight and composition; Einstein extended the principle to show that gravitation is identical (equivalent) to acceleration.

Escape velocity—the speed at which a body must travel in order to escape a strong gravitational field; rockets fired into orbits around the Earth have calculated escape velocities, as do galaxies at the periphery of galaxy clusters.

Ether (or aether)—a substance whose origins were in the Greek concept of "quintessence," the ether was the medium through which energy and matter moved, something more than a vacuum and less than air; in the late nineteenth century the Michelson-Morley experiment disproved the existence of the ether.

Euclidean geometry—plane geometry developed by the third-century BC Greek mathematician Euclid; in this geometry, parallel lines never meet.

Fermion—a particle with half-integer spin, like protons and electrons, that make up matter.

Field—a physical term describing the forces between massive bodies in gravity and electric charges in electromagnetism; Michael Faraday discovered the concept of field when studying magnetic conductors.

Field equations—differential equations describing the physical properties of interacting massive particles in gravity and electric charges in electromagnetism; Maxwell's equations for electromagnetism and Einstein's equations of gravity are prominent examples in physics.

Fifth force or **"skew" force**—a new force in MOG that has the effect of modifying gravity over limited length scales; it is carried by a particle with mass called the phion.

Fine-tuning—the unnatural cancellation of two or more large numbers involving an absurd number of decimal places, when one is attempting to explain a physical phenomenon; this signals that a true understanding of the physical phenomenon has not been achieved.

Fixed stars—an ancient Greek concept in which all the stars were static in the sky, and moved around the Earth on a distant crystalline sphere.

Frame of reference—the three spatial coordinates and one time coordinate that an observer uses to denote the position of a particle in space and time.

Galaxy—organized group of hundreds of billions of stars, such as our Milky Way.

Galaxy cluster—many galaxies held together by mutual gravity but not in as organized a fashion as stars within a single galaxy.

Galaxy rotation curve—a plot of the Doppler shift data recording the observed velocities of stars in galaxies; those stars at the periphery of giant spiral galaxies are observed to be going faster than they "should be" according to Newton's and Einstein's gravity theories.

General relativity—Einstein's revolutionary gravity theory, created in 1916 from a mathematical generalization of his theory of special relativity; it changed our concept of gravity from Newton's universal force to the warping of the geometry of spacetime in the presence of matter and energy.

Geodesic—the shortest path between two neighboring points, which is a straight line in Euclidian geometry, and a unique curved path in four-dimensional spacetime.

Globular cluster—a relatively small, dense system of up to millions of stars occurring commonly in galaxies.

Gravitational lensing—the bending of light by the curvature of spacetime; galaxies and clusters of galaxies act as lenses, distorting the images of distant bright galaxies or quasars as the light passes through or near them.

Gravitational mass—the active mass of a body that produces a gravitational force on other bodies.

Gravitational waves—ripples in the curvature of spacetime predicted by general relativity; although any accelerating body can produce gravitational radiation or waves, those that could be detected by experiments would be caused by cataclysmic cosmic events.

Graviton—the hypothetical smallest packet of gravitational energy, comparable to the photon for electromagnetic energy; the graviton has not yet been seen experimentally.

Group (in mathematics)—in abstract algebra, a set that obeys a binary operation that satisfies certain axioms; for example, the property of addition of integers makes a group; the branch of mathematics that studies groups is called group theory.

Hadron—a generic word for fermion particles that undergo strong nuclear interactions.

Hamiltonian—an alternative way of deriving the differential equations of motion for a physical system using the calculus of variations; Hamilton's principle is also called the "principle of stationary action" and was originally formulated by Sir William Rowan Hamilton for classical mechanics; the principle applies to classical fields such as the gravitational and electromagnetic fields, and has had important applications in quantum mechanics and quantum field theory.

Homogeneous—in cosmology, when the universe appears the same to all observers, no matter where they are in the universe.

Inertia—the tendency of a body to remain in uniform motion once it is moving, and to stay at rest if it is at rest; Galileo discovered the law of inertia in the early seventeenth century.

Inertial mass—the mass of a body that resists an external force; since Newton, it has been known experimentally that inertial and gravitational mass are equal; Einstein used this equivalence of inertial and gravitational mass to postulate his equivalence principle, which was a cornerstone of his gravity theory.

Inflation theory—a theory proposed by Alan Guth and others to resolve the flatness, horizon, and homogeneity problems in the standard big bang model; the very early universe is pictured as expanding exponentially fast in a fraction of a second.

Interferometry—the use of two or more telescopes, which in combination create a receiver in effect as large as the distance between them; radio astronomy in particular makes use of interferometry.

Inverse square law—discovered by Newton, based on earlier work by Kepler, this law states that the force of gravity between two massive bodies or point particles decreases as the inverse square of the distance between them.

Isotropic—in cosmology, when the universe looks the same to an observer, no matter in which direction she looks.

Kelvin temperature scale—designed by Lord Kelvin (William Thomson) in the mid-1800s to measure very cold temperatures, its starting point is absolute zero, the coldest possible temperature in the universe, corresponding to –273.15 degrees Celsius; water's freezing point is 273.15K (0°C), while its boiling point is 373.15K (100°C).

Lagrange points—discovered by the Italian-French mathematician Joseph-Louis Lagrange, these five special points are in the vicinity of two orbiting masses where a third, smaller mass can orbit at a fixed distance from the larger masses; at the Lagrange points, the gravitational pull of the two large masses precisely equals the centripetal force required to keep the third body, such as a satellite, in a bound orbit; three of the Lagrange points are unstable, two are stable.

Lagrangian—named after Joseph-Louis Lagrange, and denoted by L, this mathematical expression summarizes the dynamical properties of a physical system; it is defined in classical mechanics as the kinetic energy T minus the potential energy V; the equations of motion of a system of particles may be derived from the Euler-Lagrange equations, a family of partial differential equations.

Light cone—a mathematical means of expressing past, present, and future space and time in terms of spacetime geometry; in four-dimensional Minkowski spacetime, the light rays emanating from or arriving at an event separate spacetime into a past cone and a future cone which meet at a point corresponding to the event.

Lorentz transformations—mathematical transformations from one inertial frame of reference to another such that the laws of physics remain the same; named after Hendrik Lorentz, who developed them in 1904, these transformations form the basic mathematical equations underlying special relativity.

Mercury anomaly—a phenomenon in which the perihelion of Mercury's orbit advances more rapidly than predicted by Newton's equations of gravity; when Einstein showed that his gravity theory predicted the anomalous precession, it was the first empirical evidence that general relativity might be correct.

Meson—a short-lived boson composed of a quark and an antiquark, believed to bind protons and neutrons together in the atomic nucleus.

Metric tensor—mathematical symmetric tensor coefficients that determine the infinitesimal distance between two points in spacetime; in effect the metric tensor distinguishes between Euclidean and non-Euclidean geometry.

Michelson-Morley experiment—1887 experiment by Albert Michelson and Edward Morley that proved that the ether did not exist; beams of light traveling in the same direction, and in the perpendicular direction, as the supposed ether showed no difference in speed or arrival time at their destination.

Milky Way—the spiral galaxy that contains our solar system.

Minkowski spacetime—the geometrically flat spacetime, with no gravitational effects, first described by the Swiss mathematician Hermann Minkowski; it became the setting of Einstein's theory of gravity.

MOG—my relativistic modified theory of gravitation, which generalizes Einstein's general relativity; MOG stands for "Modified Gravity."

MOND—a modification of Newtonian gravity published by Mordehai Milgrom in 1983; this is a nonrelativistic phenomenological model used to describe rotational velocity curves of galaxies; MOND stands for "Modified Newtonian Dynamics."

Neutrino—an elementary particle with zero electric charge; very difficult to detect, it is created in radioactive decays and is able to pass through matter almost undisturbed; it is considered to have a tiny mass that has not yet been accurately measured.

Neutron—an elementary and electrically neutral particle found in the atomic nucleus, and having about the same mass as the proton.

Nuclear force—another name for the strong force that binds protons and neutrons together in the atomic nucleus.

Nucleon—a generic name for a proton or neutron within the atomic nucleus.

Neutron star—the collapsed core of a star that remains after a supernova explosion; it is extremely dense, relatively small, and composed of neutrons.

Newton's gravitational constant—the constant of proportionality, G, which occurs in the Newtonian law of gravitation, and says that the attractive force between two bodies is proportional to the product of their masses and inversely proportional to the square of the distance between them; its numerical value is:
$$G = 6.67428 \pm 0.00067 \times 10^{-11} \text{ m}^3 \text{ kg}^{-1} \text{ s}^{-2}.$$

Nonsymmetric field theory (Einstein)—a mathematical description of the geometry of spacetime based on a metric tensor that has both a symmetric part and an antisymmetric part; Einstein used this geometry to formulate a unified field theory of gravitation and electromagnetism.

Nonsymmetric Gravitation Theory (NGT)—my generalization of Einstein's purely gravitation theory (general relativity) that introduces the antisymmetric field as an extra component of the gravitational field; mathematically speaking, the nonsymmetric field structure is described by a non-Riemannian geometry.

Parallax—the apparent movement of a nearer object relative to a distant background when one views the object from two different positions; used with triangulation for measuring distances in astronomy.

Paradigm shift—a revolutionary change in belief, popularized by the philosopher Thomas Kuhn, in which the majority of scientists in a given field discard a traditional theory of nature in favor of a new one that passes the tests of experiment and observation; Darwin's theory of natural selection, Newton's gravity theory, and Einstein's general relativity all represented paradigm shifts.

Parsec—a unit of astronomical length equal to 3.262 light years.

Particle-wave duality—the fact that light in all parts of the electromagnetic spectrum (including radio waves, X-rays, etc., as well as visible light) sometimes acts like waves and sometimes acts like particles or photons; gravitation may be similar, manifesting as waves in spacetime or graviton particles.

Perihelion—the position in a planet's elliptical orbit when it is closest to the sun.

Perihelion advance—the movement, or changes, in the position of a planet's perihelion in successive revolutions of its orbit over time; the most dramatic perihelion advance is Mercury's, whose orbit traces a rosette pattern.

Perturbation theory—a mathematical method for finding an approximate solution to an equation that cannot be solved exactly, by expanding the solution in a series in which each successive term is smaller than the preceding one.

Phion—name given to the massive vector field in MOG; it is represented both by a boson particle, which carries the fifth force, and a field.

Photoelectric effect—the ejection of electrons from a metal by X-rays, which proved the existence of photons; Einstein's explanation of this effect in 1905 won him the Nobel Prize in 1921; separate experiments proving and demonstrating the existence of photons were performed in 1922 by Thomas Millikan and Arthur Compton, who received the Nobel Prize for this work in 1923 and 1927, respectively.

Photon—the quantum particle that carries the energy of electromagnetic waves; the spin of the photon is 1 times Planck's constant h.

Pioneer 10 and 11 spacecraft—launched by NASA in the early 1970s to explore the outer solar system, these spacecraft show a small, anomalous acceleration as they leave the inner solar system.

Planck's constant (h)—a fundamental constant that plays a crucial role in quantum mechanics, determining the size of quantum packages of energy such as the photon; it is named after Max Planck, a founder of quantum mechanics; the symbol $\hbar$, which equals h divided by 2π, is used in quantum mechanical calculations.

Principle of general covariance—Einstein's principle that the laws of physics remain the same whatever the frame of reference an observer uses to measure physical quantities.

Principle of least action—more accurately the principle of *stationary* action, this variational principle, when applied to a mechanical system or a field theory, can be used to derive the equations of motion of the system; the credit for discovering the principle is given to Pierre-Louis Moreau Maupertius but it may have been discovered independently by Leonhard Euler or Gottfried Leibniz.

Proton—an elementary particle that carries positive electrical charge and is the nucleus of a hydrogen atom.

Ptolemaic model of the universe—the predominant theory of the universe until the Renaissance, in which the Earth was the heavy center of the universe and all other heavenly bodies, including the moon, sun, planets, and stars, orbited around it; named for Claudius Ptolemy.

Quantize—to apply the principles of quantum mechanics to the behavior of matter and energy (such as the electromagnetic or gravitational field energy); breaking down a field into its smallest units or packets of energy.

Quantum field theory—the modern relativistic version of quantum mechanics used to describe the physics of elementary particles; it can also be used in nonrelativistic fieldlike systems in condensed matter physics.

Quantum gravity—attempts to unify gravity with quantum mechanics.

Quantum mechanics—the theory of the interaction between quanta (radiation) and matter; the effects of quantum mechanics become observable at the submicroscopic distance scales of atomic and particle physics, but macroscopic quantum effects can be seen in the phenomenon of quantum entanglement.

Quantum spin—the intrinsic quantum angular momentum of an elementary particle; this is in contrast to the classical orbital angular momentum of a body rotating about a point in space.

Quark—the fundamental constituent of all particles that interact through the strong nuclear force; quarks are fractionally charged, and come in several varieties; because they are confined within particles such as protons and neutrons, they cannot be detected as free particles.

Quasars—"quasi-stellar objects," the farthest distant objects that can be detected with radio and optical telescopes; they are exceedingly bright, and are believed to be young, newly forming galaxies; it was the discovery of quasars in 1960 that quashed the steady-state theory of the universe in favor of the big bang.

Quintessence—a fifth element in the ancient Greek worldview, along with earth, water, fire and air, whose purpose was to move the crystalline spheres that supported the heavenly bodies orbiting the Earth; eventually this concept became known as the "ether," which provided the *something* that bodies needed to be in contact with in order to move; although Einstein's special theory of relativity dispensed with the ether, recent explanations of the acceleration of the universe call the varying negative pressure vacuum energy "quintessence."

Redshift—a useful phenomenon based on the Doppler principle that can indicate whether and how fast bodies in the universe are receding from an observer's position on Earth; as galaxies move rapidly away from us, the frequency of the wavelength of their light is shifted toward the red end of the electromagnetic spectrum; the amount of this shifting indicates the distance of the galaxy.

Riemann curvature tensor—a mathematical term that specifies the curvature of four-dimensional spacetime.

Riemannian geometry—a non-Euclidean geometry developed in the mid-nineteenth century by the German mathematician George Bernhard Riemann that describes curved surfaces on which parallel lines *can* converge, diverge, and even intersect, unlike Euclidean geometry; Einstein made Riemannian geometry the mathematical formalism of general relativity.

Satellite galaxy—a galaxy that orbits a host galaxy or even a cluster of galaxies.

Scalar field—a physical term that associates a value without direction to every point in space, such as temperature, density, and pressure; this is in contrast to a vector field, which has a direction in space; in Newtonian physics or in electrostatics, the potential energy is a scalar field and its gradient is the vector force field; in quantum field theory, a scalar field describes a boson particle with spin zero.

Scale invariance—distribution of objects or patterns such that the same shapes and distributions remain if one increases or decreases the size of the length scales or space in which the objects are observed; a common example of scale invariance is fractal patterns.

Schwarzschild solution—an exact spherically symmetric static solution of Einstein's field equations in general relativity, worked out by the astronomer Karl Schwarzschild in 1916, which predicted the existence of black holes.

Self-gravitating system—a group of objects or astrophysical bodies held together by mutual gravitation, such as a cluster of galaxies; this is in contrast to a "bound system" like our solar system, in which bodies are mainly attracted to and revolve around a central mass.

Singularity—a place where the solutions of differential equations break down; a spacetime singularity is a position in space where quantities used to determine the gravitational field become infinite; such quantities include the curvature of spacetime and the density of matter.

Spacetime—in relativity theory, a combination of the three dimensions of space with time into a four-dimensional geometry; first introduced into relativity by Hermann Minkowski in 1908.

Special theory of relativity—Einstein's initial theory of relativity, published in 1905, in which he explored the "special" case of transforming the laws of physics from one uniformly moving frame of reference to another; the equations of special relativity revealed that the speed of light is a constant, that objects appear contracted in the direction of motion when moving at close to the speed of light, and that $E = mc^2$, or energy is equal to mass times the speed of light squared.

Spin—see quantum spin.

String theory—a theory based on the idea that the smallest units of matter are not point particles but vibrating strings; a popular research pursuit in physics for two decades, string theory has some attractive mathematical features, but has yet to make a testable prediction.

Strong force—see nuclear force.

Supernova—spectacular, brilliant death of a star by explosion and the release of heavy elements into space; supernovae type 1a are assumed to have the same intrinsic brightness and are therefore used as standard candles in estimating cosmic distances.

Supersymmetry—a theory developed in the 1970s which, proponents claim, describes the most fundamental spacetime symmetry of particle physics: For every boson particle there is a supersymmetric fermion partner, and for every fermion there exists a supersymmetric boson partner; to date, no supersymmetric particle partner has been detected.

Tully-Fisher law—a relation stating that the asymptotically flat rotational velocity of a star in a galaxy, raised to the fourth power, is proportional to the mass or luminosity of the galaxy.

Unified theory (or unified field theory)—a theory that unites the forces of nature; in Einstein's day those forces consisted of electromagnetism and gravity; today the weak and strong nuclear forces must also be taken into account, and perhaps someday MOG's fifth force or skew force will be included; no one has yet discovered a successful unified theory.

Vacuum—in quantum mechanics, the lowest energy state, which corresponds to the vacuum state of particle physics; the vacuum in modern quantum field theory is the state of perfect balance of creation and annihilation of particles and antiparticles.

Variable Speed of Light (VSL) cosmology—an alternative to inflation theory, which I proposed in 1993, in which the speed of light was much faster at the beginning of the universe than it is today; like inflation, this theory solves the horizon and flatness problems in the very early universe in the standard big bang model.

Vector field—a physical value that assigns a field with the position and direction of a vector in space; it describes the force field of gravity or the electric and magnetic force fields in James Clerk Maxwell's field equations.

Virial theorem—a means of estimating the average speed of galaxies within galaxy clusters from their estimated average kinetic and potential energies.

Vulcan—a hypothetical planet predicted by the nineteenth-century astronomer Urbain Jean Joseph Le Verrier to be the closest orbiting planet to the sun; the presence of Vulcan would explain the anomalous precession of the perihelion of Mercury's orbit; Einstein later explained the anomalous precession in general relativity by gravity alone.

Weak force—one of the four fundamental forces of nature, associated with radioactivity such as beta decay in subatomic physics; it is much weaker than the strong nuclear force but still much stronger than gravity.

X-ray clusters—galaxy clusters with large amounts of extremely hot gas within them that emit X-rays; in such clusters, this hot gas represents at least twice the mass of the luminous stars.

Bibliography

Barrow, John D., and Frank J. Tipler. *The Anthropic Cosmological Principle.* Oxford, UK: Oxford University Press, 1986.

Baum, Richard, and William Sheehan. *In Search of Planet Vulcan: The Ghost in Newton's Clockwork Universe.* Cambridge, MA: Perseus, Basic Books, 1997.

Baumann, Mary K., Will Hopkins, Loralee Nolletti, and Michael Soluri. *What's Out There: Images from Here to the Edge of the Universe.* London: Duncan Baird Publishers, 2005.

Bolles, Edmund Blair. *Einstein Defiant: Genius versus Genius in the Quantum Revolution.* Washington, DC: John Henry Press, 2004.

Boorstin, Daniel J. *The Discoverers: A History of Man's Search to Know His World and Himself.* New York: Random House, 1983.

Brown, Harvey R. *Physical Relativity: Space-time Structure from a Dynamical Perspective.* Oxford, UK: Clarendon Press, 2005.

Christianson, Gale E. *In the Presence of the Creator: Isaac Newton & His Times.* New York, London: The Free Press, Division of Macmillan, Inc., 1984.

Clark, Ronald W. *Einstein: The Life and Times.* New York: World Publishing, 1971.

Coveney, Peter and Roger Highfield. *The Arrow of Time: A Voyage through Science to Solve Time's Greatest Mystery.* New York: Fawcett Columbine, 1990.

Crelinsten, Jeffrey. *Einstein's Jury: The Race to Test Relativity.* Princeton, N.J.: Princeton University Press, 2006.

Darling, David. *Gravity's Arc: The Story of Gravity, from Aristotle to Einstein and Beyond.* Hoboken: John Wiley & Sons, 2006.

Davies, P.C.W. *Space and Time in the Modern Universe.* Cambridge, London, New York, Melbourne: Cambridge University Press, 1977.

———. *The Physics of Time Asymmetry.* Berkeley and Los Angeles: University of California Press, 2nd edition, 1977.

Eisenstaedt, Jean. *The Curious History of Relativity: How Einstein's Theory of Gravity Was Lost and Found Again.* Princeton, N.J.: Princeton University Press, 2006.

Farrell, John. *The Day Without Yesterday: Lemaître, Einstein, and the Birth of Modern Cosmology*. New York: Thunder's Mouth Press, Avalon Publishing Group, 2005.

Ferguson, Kitty. *Tycho & Kepler: The Unlikely Partnership that Forever Changed Our Understanding of the Heavens*. New York: Walker and Company, 2002.

Freeman, Ken, and Geoff McNamara. *In Search of Dark Matter*. Chichester, UK: Praxis Publishing Ltd., 2006.

Gleick, James. *Isaac Newton*. New York: Pantheon Books, 2003.

Gondhalekar, Prabhakar. *The Grip of Gravity: The Quest to Understand the Laws of Motion and Gravitation*. Cambridge: Cambridge University Press, 2001.

Greene, Brian. *The Elegant Universe: Superstrings, Hidden Dimensions, and the Quest for the Ultimate Theory*. New York and London: W.W. Norton & Company, 1999.

Grosser, Morton. *The Discovery of Neptune*. New York: Dover Publications, Inc., 1979.

Guth, Alan H. *The Inflationary Universe: The Quest for a New Theory of Cosmic Origins*. Cambridge, MA: Perseus Books, 1997.

Highfield, Roger, and Paul Carter. *The Private Lives of Albert Einstein*. London, Boston: Faber and Faber, 1993.

Isaacson, Walter. *Einstein: His Life and Universe*. New York, London, Toronto, Sydney: Simon & Schuster, 2007.

Johnson, George. *Miss Leavitt's Stars: The Untold Story of the Woman Who Discovered How to Measure the Universe*. New York, London: Atlas Books, W.W. Norton & Company, 2005.

Kirshner, Robert P. *The Extravagant Universe: Exploding Stars, Dark Energy and the Accelerating Cosmos*. Princeton and Oxford: Princeton University Press, 2002.

Krauss, Lawrence. *Quintessence: The Mystery of Missing Mass in the Universe*. New York: Basic Books, Perseus, 2000.

Kuhn, Thomas S. *The Structure of Scientific Revolutions*. Chicago and London: The University of Chicago Press, third edition, 1996.

Kuhn, Thomas S., James Conant, and John Haugeland, eds. *The Road Since Structure*. Chicago and London: The University of Chicago Press, 2000.

Lightman, Alan. *The Discoveries: Great Breakthroughs in Twentieth-Century Science*. Toronto: Alfred A. Knopf Canada, 2005.

Livio, Mario. *The Accelerating Universe: Infinite Expansion, the Cosmological Constant, and the Beauty of the Cosmos*. New York, Chichester, Weinheim, Brisbane, Singapore, Toronto: John Wiley & Sons, Inc., 2000.

Lorentz, H. A., A. Einstein, H. Minkowski, and H. Weyl. *The Principle of Relativity: A Collection of Original Memoirs on the Special and General Theory of Relativity*. London: Methuen & Co. Ltd., 1923.

Magueijo, João. *Faster than the Speed of Light: The Story of a Scientific Speculation*. Cambridge, MA: Perseus Publishing, 2003.

Mather, John C., and John Boslough. *The Very First Light: The True Inside Story of the Scientific Journey Back to the Dawn of the Universe.* New York: Basic Books, 1996.

Pais, Abraham. *"Subtle is the Lord": The Science and the Life of Albert Einstein.* New York: Oxford University Press, 1982.

Poincaré, Henri, and Francis Maitland, translator. *Science and Method.* Mineola, NY: Dover Publications, 2003.

Rees, Martin. *Our Cosmic Habitat.* London: Phoenix, Orion Books, 2003.

Savitt, Steven T., ed. *Time's Arrows Today: Recent Physical and Philosophical Work on the Direction of Time.* Cambridge, UK: Cambridge University Press, 1997.

Schilpp, Paul Arthur, ed. *Albert Einstein: Philosopher-Scientist.* New York: Tudor Publishing Company, 1957.

Schutz, Bernard. *Gravity from the Ground Up: An Introductory Guide to Gravity and General Relativity.* Cambridge, UK: Cambridge University Press, 2003.

Silk, Joseph. *The Big Bang: The Creation and Evolution of the Universe.* San Francisco: W.H. Freeman and Company, 1980.

Singh, Simon. *Big Bang: The Origin of the Universe.* London, New York: Fourth Estate, 2004.

Smolin, Lee. *The Trouble with Physics: The Rise of String Theory, the Fall of Science, and What Comes Next.* Boston, New York: Houghton Mifflin Company, 2006.

Steinhardt, Paul J., and Neil Turok. *Endless Universe: Beyond the Big Bang.* New York: Doubleday, 2007.

Susskind, Leonard. *The Cosmic Landscape: String Theory and the Illusion of Intelligent Design.* New York, Boston: Little, Brown and Company, 2006.

Thorne, Kip S. *Black Holes & Time Warps: Einstein's Outrageous Legacy.* New York and London: W.W. Norton & Company, 1994.

Thorne, Kip S., Charles W. Misner, and John Archibald Wheeler. *Gravitation.* New York: W.H. Freeman, 1973.

Vollmann, William T. *Uncentering the Earth: Copernicus and the Revolutions of the Heavenly Spheres.* New York, London: Atlas Books, W.W. Norton & Company, 2006.

Wali, Kameshwar C. *Chandra: A Biography of S. Chandrasekhar.* Chicago: University of Chicago Press, 1992.

Weinberg, Steven. *Gravitation and Cosmology: Principles and Applications of the General Theory of Relativity.* New York, Chichester, Brisbane, Toronto, Singapore: John Wiley & Sons, 1972.

———. *The First Three Minutes: A Modern View of the Origin of the Universe.* New York: Basic Books, 1977.

Whittaker, Sir Edmund. *A History of the Theories of Aether and Electricity: The Modern Theories 1900–1926.* London, Edinburgh, Paris, Melbourne, Toronto and New York: Thomas Nelson and Sons Ltd., 1953.

Woit, Peter. *Not Even Wrong: The Failure of String Theory and the Continuing Challenge to Unify the Laws of Physics.* London: Jonathan Cape, 2006.

Acknowledgments

This book would never have been completed without the patience and dedication of my wife, Patricia. She performed the wonderful and difficult task of editing major parts of the book and helped in researching many details necessary to complete it. I wish to thank several colleagues for their help and extensive comments on the manuscript, including Martin Green, Pierre Savaria, and my collaborator, Viktor Toth. I also thank my colleagues Paul Steinhardt, João Magueijo, Philip Mannheim, Harvey Brown, Paul Frampton, Stacy McGaugh, and Lee Smolin for helpful comments. I particularly thank my sister-in-law, Keeva Kroll, for a careful reading of the manuscript.

Many graduate students have contributed over the years to developing my modified gravity theory. Special thanks go to Joel Brownstein for many helpful discussions and all the work he did on the consequences of the theory.

I also wish to thank my editors, TJ Kelleher and Elisabeth Dyssegaard at HarperCollins (Smithsonian Books) in New York, and Janice Zawerbny at Thomas Allen in Toronto, as well as Patrick Crean and Jim Allen for their enthusiasm and support. I thank my literary agent, Jodie Rhodes, for her excitement about the book from the beginning, and for her dedication and hard work in finding the right publishers.

Finally, I thank our family for their patience, love, and support during the three years of working on this book.

Index